FRUIT& VEGETABLE QUALITY

An Integrated View

FRUIT&
VEGETABLE
QUALITY

An Integrated View

Edited by

Robert L. Shewfelt
University of Georgia, Athens

Bernhard Brückner
Institute for Vegetable and Ornamental Crops
Großbeeren, Germany

CRC Press
Taylor & Francis Group
Boca Raton London New York

CRC Press is an imprint of the
Taylor & Francis Group, an **informa** business

CRC Press
Taylor & Francis Group
6000 Broken Sound Parkway NW, Suite 300
Boca Raton, FL 33487-2742

First issued in paperback 2019

© 2000 by Taylor & Francis Group, LLC
CRC Press is an imprint of Taylor & Francis Group, an Informa business

No claim to original U.S. Government works

ISBN-13: 978-1-56676-785-9 (hbk)
ISBN-13: 978-0-367-39874-3 (pbk)

Visit the Taylor & Francis Web site at
http://www.taylorandfrancis.com

and the CRC Press Web site at
http://www.crcpress.com

CONTENTS

11. HOUSE OF QUALITY—AN INTEGRATED VIEW OF FRUIT AND VEGETABLE QUALITY   199
ANNE C. BECH

SECTION FOUR
AN INTEGRATED VIEW   225

12. ECONOMICS OF QUALITY   227
WOJCIECH J. FLORKOWSKI

13. INTEGRATED QUALITY MANAGEMENT APPLIED TO THE PROCESSED-VEGETABLES INDUSTRY   246
JACQUES VIAENE, XAVIER GELLYNCK, AND WIM VERBEKE

14. METHODS AND EXAMPLES OF INTEGRATION   267
STANLEY E. PRUSSIA

15. INTEGRATING PROBLEM-ORIENTED RESEARCH   285
ROLF KUCHENBUCH, BERNHARD BRÜCKNER,
JÖRG RÜHLMANN, AND BARBARA RÖGER

16. A MORE INTEGRATED VIEW   296
ROBERT L. SHEWFELT AND LEOPOLD M. M. TIJSKENS

IN May of 1997, more than 50 scientists from 15 countries met in Potsdam, Germany, to discuss the development of a more integrated view of fruit and vegetable quality. The purpose of the conference was to bring together leading scientists from different disciplines, expose them to different perspectives, and develop a framework for future integration. To the surprise of the organizers of the conference, each of the speakers appeared to have an appreciation for the need of integration and presented evidence that their research was incorporating some degree of integration. Despite such promising signs, actual progress will be limited by the lack of any agreement as to where to start or even what the term "integration" really means.

This book represents the vision that emerged from that conference. It is designed for anyone involved in management, production, handling, distribution or processing of fruits and vegetables who believes that improved quality requires integration across business functions and scientific disciplines. It provides concise descriptions of important issues facing postharvest handlers, roadmaps to the literature in specific fields, assessments of current knowledge and research needs, and specific examples of product-based research. It provides a guide to the dynamic developments in integrating projects on fruit and vegetable quality and a range of options to achieve desired effects of better coordination of research across scientific disciplines and to improve postharvest handling to enhance fruit and vegetable quality.

Lack of a common vocabulary for quality, acceptability and related terms as well as a clear need for measures that can be used across traditional disciplinary lines and permit comparison of similar studies were identified as the primary factors limiting integration. The underlying premise of this book is that a greater emphasis on collaborative research that crosses interdisciplinary lines is more likely to lead to improved fruit and vegetable quality than a continued emphasis on rigorous, single-disciplinary studies. *Fruit and Vegetable Quality: An Integrated View* presents 15 unique perspectives on the topic culminating in a final chapter that seeks a common ground for bringing together these perspectives into a unified language that will permit a more integrated approach to fruit and vegetable research and postharvest handling. The book seeks standardization of common terms and mathematical expression while remaining flexible enough to permit innovation in a dynamic field of study.

Section I introduces the aspects of fruits and vegetables that have captivated the attention of consumers and the press around the world— handling and distribution to preserve freshness and ways to breed crops for specific quality characteristics. This section emphasizes those aspects of fruit and vegetable quality that have focused the spotlight on the topic of what is driving postharvest research and provide the rationale for greater integration. Section II provides an appreciation of cultural, environmental, handling and storage techniques that are available to handlers and distributors today. This area is where the greatest improvements of quality have been made to date, but future advances here will probably have limited impact without integration both with crop production and a better understanding of consumer behavior. Section III provides a perspective on fruit and vegetable quality, bringing in the consumer who is the ultimate judge of quality. Section IV provides four perspectives on how studies have been integrated across disciplines in the past, with a concluding chapter that attempts to make projections into the future to provide a basis for a more integrated approach to fruit and vegetable quality. At the beginning and the end of each section, we have assessed the current situation and suggested future avenues of research. The success of this book will rest with its ability to stimulate greater research cooperation across disciplinary perspectives leading to improved quality of fresh fruits and vegetables available to the consumer.

ROBERT L. SHEWFELT
BERNHARD BRÜCKNER

Zainul Andani, Ph.D.
AFRC Institute of Food
  Research
Whiteknights Road, Earley
  Gate
Reading RG6 2EF, United
  Kingdom

Helga Auerswald, Ph.D.
Institute for Vegetable and
  Ornamental Crops
Thedor-Echtermeyer-Weg 1
14979 Grossbeeren, Germany

Anne C. Bech, Ph.D.
The MAPP Centre
Aarhus School of Business and
  Administration
Fuglesangs Alle 4
8210 Aarhus V, Denmark

Bernhard Brückner, Ph.D.
Institute for Vegetable and
  Ornamental Crops

Thedor-Echtermeyer-Weg 1
14979 Grossbeeren, Germany

Wojciech J. Florkowski, Ph.D.
Department of Agricultural and
  Applied Economics
University of Georgia
Griffin, GA 30223, USA

Xavier Gellynck, Ph.D.
University of Ghent
Department of Agricultural
  Economics–Agro-Marketing
Coupure Links 653
9000 Gent, Belgium

Susanne Huyskens-Keil, Ph.D.
Humboldt University Berlin
Königin-Luise-Str. 22
14195 Berlin, Germany

David S. Johnson, Ph.D.
East Malling Research Station

Fruit Storage Division
East Malling/West Malling
Kent MEi 96BT, United
  Kingdom

Wim M. F. Jongen, Ph.D.
Wageningen Agricultural
  University
Department of Food
  Technology and Nutritional
  Sciences
Bomenweg 2
7603 HD Wageningen,
  Netherlands

Angelika Krumbein, Ph.D.
Institute for Vegetable and
  Ornamental Crops
Thedor-Echtermeyer-Weg 1
14979 Grossbeeren, Germany

Rolf Kuchenbuch, Ph.D.
Institute for Vegetable and
  Ornamental Crops
Thedor-Echtermeyer-Weg 1
14979 Grossbeeren, Germany

Manfred Linke, Ph.D.
Institute of Agricultural
  Engineering
Max-Eyth-Allee 100
14469 Potsdam-Bornim,
  Germany

Hal J. H. MacFie, Ph.D.
AFRC Institute of Food
  Research
Whiteknights Road, Earley Gate
Reading RG6 2EF, United
  Kingdom

Torsten Nilsson, Ph.D.
Swedish Agricultural University
Department of Horticulture
P.O. Box 55
S-230 53 Alnarp, Sweden

Matthais von Oppen, Ph.D.
University Hohenheim
Intsitut für Agrar-und
  Sozialökonomie in den
Tropen und Subtropen
Schloss Osthof-Süd
70599 Stuttgart, Germany

Susanne Pecher, Ph.D.
University Hohenheim
Intsitut für Agrar-und
  Sozialökonomie in den
Tropen und Subtropen
Schloss Osthof-Sud
70599 Stuttgart, Germany

Stanley E. Prussia, Ph.D.
Department of Agricultural and
  Biological Sciences
University of Georgia
  Experiment Station
Griffin, GA 30223, USA

Martin S. Ridout, Ph.D.
East Malling Research Station
Fruit Storage Division
East Malling/West Malling
Kent MEi 96BT, United
  Kingdom

Barbara Röger, M.S.
Institute for Vegetable and
  Ornamental Crops

Thedor-Echtermeyer-Weg 1
14979 Grossbeeren, Germany

Jörg Rühlmann, Ph.D.
Institute for Vegetable and
    Ornamental Crops
Thedor-Echtermeyer-Weg 1
14979 Grossbeeren, Germany

Ilona Schonhof, Ph.D.
Institute for Vegetable and
    Ornamental Crops
Thedor-Echtermeyer-Weg 1
14979 Grossbeeren, Germany

Monika Schreiner, Ph.D.
Institute for Vegetable and
    Ornamental Crops
Thedor-Echtermeyer-Weg 1
14979 Grossbeeren, Germany

Robert L. Shewfelt, Ph.D.
Department of Food Science
    and Technology
University of Georgia
Athens, GA 30602, USA

Leopold M. M. Tijskens
ATO-DLO

P.O. Box 17
6700 AA Wageningen,
    Netherlands

Wim Verbeke
University of Ghent
Department of Agricultural
    Economics–Agro-Marketing
Coupure Links 653
9000 Gent, Belgium

Jacques Viaene, Ph.D.
University of Ghent
Department of Agricultural
    Economics–Agro-Marketing
Coupure Links 653
9000 Gent, Belgium

Peter Wehling, Ph.D.
Bundesanstalt fur
    Züchtungsforschung an
    Kulturpflanzen
Institut für Züchtungmethodik
    landwirtschaftlicher
Kulturpflanzen
Institutsplatz 1
18190 Gross Lüsewitz,
    Germany

CONTEMPORARY ISSUES

CONTEXT

- Marketing and distribution of fresh fruit and vegetables encompasses a wide range of operations and activities.
- Commercial cultivars, selected for their ability to withstand the rigors of marketing and distribution, tend to lack sufficient quality, particularly flavor.
- Quantitative models that clearly link quality characteristics to economic value or selection traits are not available.

OBJECTIVES

- To present the supply chain as an integrating framework for research on issues of fresh fruit and vegetable quality.
- To describe the potential for improved quality using genetic engineering.
- To relate economic principles of consumer preference to those of plant breeding and selection.

CHAPTER 1

Food Supply Chains: From Productivity toward Quality

WIM M. F. JONGEN

INTRODUCTION

ALTHOUGH the phrase food supply chain has been widely used, there is no accurate definition of the term. In this introductory chapter an approach will be used that contains two primary elements. It uses a consumer-oriented approach and focuses on product flows between companies in a specific supply chain. Traditionally, food supply chains have been characterized by two distinct features (Jongen, 1996):

1. One-way communication through the chain from producers of raw materials and/or primary products to the users of end-products
2. A poor understanding of the concept of product quality, which is still predominantly based on technical criteria and producer-focus particularly with respect to costs and productivity

In a growing market as experienced for some decades in most West-European countries, this approach has been very successful. However, in the last decade the market situation has changed dramatically. In addition to market saturation resulting from steadily increasing production levels, other developments have had a large influence on the market. For example, the size of the market is limited by the reduction in the rate of population growth, and, in a number of countries, the saturation point for caloric intake has been reached. Other important changes include demographic shifts toward an aging population and major changes in household composition (Meulenberg and Viaene, 1998).

Generally, the consumer is better educated and expects a wider range of products. Further, consumers are less predictable in their purchase behavior, eat more outside homes and are more conscious about health-related aspects of fresh fruits and vegetables (Popcorn, 1992). Also the perception of product quality is no longer restricted to the physical properties of the product but also other agents including production practices. In addition, scientific progress such as developments in biotechnology or new processing technologies will result in product innovation while improved analytical methodology will lead to demands for improved product safety and quality.

As a result, there is a continual need for new products and a more differentiated food product assortment. Related to this development, product life cycles are becoming shorter (Meulenberg, 1997), hence the efficiency and flexibility of food production systems become even more important. Currently the development and introduction of new food products requires large investments in money and labor with only a relatively small chance of success. Fuller (1994) has estimated that the period from idea until product introduction is roughly between 4–7 years. Fifty percent of the new products will not survive the test market phase, more than 20% of newly launched products will disappear from the market within three years and after five years less than 20% is still on the market. Thus, product innovation is still primarily a matter of trial and error.

The history of product development shows that over successive generations of products, quality demands increase in complexity. For first-generation products, storability and the prolongation of shelf life and prevention of spoilage were the most important criteria. In the second generation, the need to meet good nutritional standards is included. Flavor and convenience characterized the third generation while the fourth also carried the health-promoting capacity as a quality attribute.

The interdependency between consumer desires, on the one hand, and development of technologies and research on the other hand has been recognized by many food-producing companies but it has not been implemented systematically yet and requires more attention. This chapter will focus on these interrelationships and a more systematic means of integrating them into the supply chain.

CHAIN REVERSAL AS A STARTING POINT

The many changes in the market for both fresh and processed food products call for a repositioning of existing food production systems and raise the question whether the concepts currently used can survive the challenges of the future. The large number of changes and the costs as-

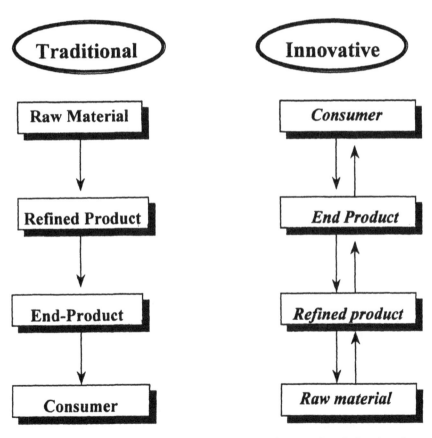

FIGURE 1.1 Schematic picture of the traditional and innovative chain-oriented approach.

sociated with product innovation make it necessary to develop a new approach toward the production of food products, namely, chain reversal in which the consumer has become the starting point of thinking. This reversal requires thorough knowledge of the market and consumer preferences to provide quality control throughout the chain (Meulenberg, 1997). A definition of product quality is the first step in this new approach. It requires translation of consumer perceptions and preferences into product characteristics that are measurable and can be specified. Using this concept raw material composition and properties can be coupled directly to the quality of the end product. A distinction can be made between the possibilities of achieving the desired end-product quality via modulation of processing conditions and optimization of raw material properties, e.g., via biotechnology (Jongen, 1996). In Figure 1.1 a schematic picture is drawn of the traditional (A) and the innovative (B) chain-oriented approach is presented.

PRODUCT QUALITY

Development of a definition of the term "product quality" has been the subject of a large number of studies and the contributions of Juran, Deming, Crosby and others to our thinking about quality are invaluable. Juran (1989) included the consumer in his concept of product quality and prevention of defects instead of control of quality. His definition is "Quality is fitness for use." With respect to food products, a better wording would be "Quality is to meet the expectations of the consumer." Two important aspects should be emphasized: (1) the consumer is the starting point of thinking about quality and (2) the consumer does not work with accurate specifications. *The* consumer doesn't exist. There is no average consumer. Instead, there is a specific consumer, who, in a specific situation and at a certain moment, has a specific need to which the producer can respond. The consumer buys and consumes a product for a number of reasons. Partly these reasons refer to product properties, partly they refer to the production system.

I propose to use here the terms intrinsic and extrinsic factors. "Intrinsic factors" refer to physical properties such as flavor, texture, appearance, shelf life and nutritional value. These properties are directly measurable and objective. In this context a food product does not possess quality as such but its physical properties, which are turned into quality attributes by the perception of the consumer. An example is the texture of an apple. From a physicochemical point of view, the texture of a product can be described in terms of cell wall composition and structure. Consumers perceive texture during consumption and describe their experience in terms of crispiness or mealiness and toughness. The total of quality attributes (intrinsic factors) determines the quality of a given product. "Extrinsic factors" refer to the production system and include factors such as the amount of pesticides used during growing, the type of packaging material used, a specific processing technology or the use

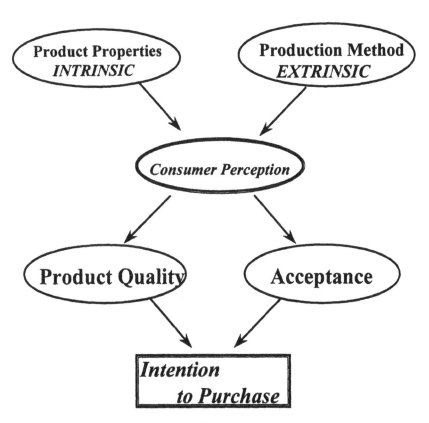

FIGURE 1.2 The concept of the total of intrinsic and extrinsic factors determining purchase behavior.

of biotechnology to modify product properties. Extrinsic factors do not necessarily have a direct influence on physical properties but influence the acceptance of the product for consumers. The total of intrinsic and extrinsic factors determines the purchase behavior (Jongen, 1995). In Figure 1.2 this concept is visualized.

Quality Dimensions in the Production-Distribution Cycle

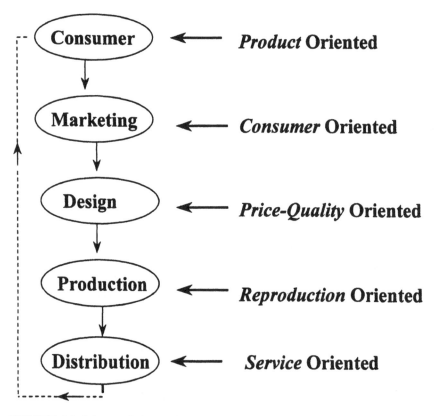

FIGURE 1.3 Scheme of the production-distribution cycle indicating the quality dimensions of the actors in the chain.

Quality management can be a confusing concept, partly because people view quality relative to their role in the production chain, and there is no uniformly accepted definition of quality. The question is then which criteria and/or variables provide an adequate description of quality for a specific actor in the chain. Figure 1.3 depicts a scheme representing the production-distribution cycle, indicating the key steps in innovation and production (Jongen, 1996b).

The consumer is the driving force for product innovation and the consumer generally views quality from the product-based perspective as described above. Indeed, products should meet consumer needs, and it is the task of the marketing department to identify the needs specified for the type of products that can be produced. A product that meets consumer expectations can rightly be described as having good "quality" and therefore marketing people, e.g., within a breeding company work with a user-based definition of quality. The next step is to translate consumer expectations into product characteristics and handling specifications. Product specifications include size, form finish, flavor, safety, shelf life and nutritional value. Process specifications include the type of equipment needed and handling conditions required. Product designers must balance price and quality to meet marketing objectives. Therefore the focus here is primarily oriented at price and quality. One of the special features of agricultural products is the large variation in product quality that can occur as a consequence of seasonal variations, cultivation practices and cultivar differences. Nevertheless, the producer must adhere to the design specifications, so the quality focus is reproduction-oriented. The production-distribution cycle is completed when a product has been moved via wholesale and retail outlets to the consumer. Distribution does not end the relationship between the consumer and the producer, so service-oriented concepts of quality cannot be ignored by the producer.

QUALITY AS A STEERING FACTOR FOR INNOVATION IN THE SUPPLY CHAIN

Production System Characteristics

Product innovation for fresh fruits and vegetables encompasses a number of characteristics that are specific for food products and that complicate quality control efforts throughout the chain.

1. Food products are perishable and storage conditions affect product quality and the rate of spoilage, requiring strict adherence to quality guidelines for production, storage and residence time throughout the chain.
2. Production and harvest of plant foods is seasonal while consumers expect constant availability thereby placing increased demands on storage facilities and transport technologies.
3. There is a clear awareness among consumers of the relationship between diet and health, which includes both the absence of unwanted components such as pathogens and toxicants, either naturally present or added, and ensuring the levels of components that are wanted such as vitamins and minerals. Adequate risk prevention requires systematic quality control.
4. Small-scale production is an obvious characteristic of fruits and vegetables and a complicating factor in ensuring homogeneous products, e.g., size and color of apples, requiring specialized organizational structures.
5. The large number of retail outlets and wholesale distributors increases the difficulty of adequate control of distribution quality and service orientation while making it virtually impossible to define accurate specifications for a given product.
6. Fresh food products are perishable and have only a very limited shelf life which, when combined with small store inventories to reduce costs, has resulted in systems with high delivery frequencies.
7. The primary aim of the food industry is to provide the community with food products of good quality for a reasonable price. So generally, food products have low added value, which is a major drawback for product innovation. This is particularly the case for fresh products.

Chain Ruptures and Product Quality

One characteristic of the traditional supply chain is the use of differing concepts of product quality by actors at different points in the chain. Figure 1.4 gives an example of the large number of quality criteria a plant food product has to meet and how they differ in the chain from the breeder to the consumer. It should be obvious that no single product can meet all the requirements indicated and that there is a clear need for one integrated concept of product quality throughout any specific supply chain.

Actors in the Chain and Quality Perception

Breeder   Vitality Seeds, Yield

Grower   Productivity, Uniformity,
                   Disease resistance

Auction   Uniformity, Reliable Supply,
                   Constant Quality

Distribution   Keepability, Availability,
                   Damage Sensitivity

Retailer   Good Shelf-life, Diversity,
                   Appearance, Low waste

Consumer   Tasty, healthy, Sustainable
                   Convenience, Constant Quality

FIGURE 1.4 Examples of the quality criteria used by the actors in the chain.

Another consideration for fruits and vegetables is that prior to harvest during growth and development product quality is built up whereas storage and handling inevitably lead to quality loss. This means that differing concepts have to be applied. During growth, knowledge about the relationships between agronomical conditions, cultivar and harvest is critical to achieve a good shelf life during storage and handling for fresh products. An example comes from the flavonoids, a group of secondary plant metabolites that occur in a large variety of fruits and vegetables. From a quality perspective the presence of several representatives of this group is of great importance. In apples, for example, they are responsible for color formation (anthocyanins). At the same time they influence the flavor of the products (catechins) and are of great interest for their capacity to act as an antioxidant (flavonols). The biosynthesis process, and as a consequence the levels of the various flavonoids, is dependent on a large number of factors such as cultivar differences, agronomic and storage conditions and conditions in the retail market. To enable the control of desired levels of specific flavonoids, fast and objective analytical measurement systems are required. Quality loss is inevitable because harvesting imposes stress and the postharvest conditions are vital for maintenance of good end-product quality. Fruit and vegetables are living organisms that respire and are metabolically active. The composition and structure of the primary product and the associated handling properties are strongly dependent upon metabolic processes. For a chain-oriented system of quality control, reliable methods are needed that can measure external and internal quality and predict shelf life.

Also, methods that can relate raw material composition and end-product quality to the processing conditions are necessary (Jongen, 1996). An excellent example comes from the recent work of van Kooten et al. (1997) on cucumber. Using the efficiency of the photosynthesis system as a parameter, they have developed a measurement system that can be used to measure temperature failure during storage and predict shelf life in the chain.

Cycles of Change

The market for food products of plant origin has become increasingly competitive and changing consumer demands have provided incentives for companies to innovate. One example comes from the National Food Surveillance study that was carried out in the Netherlands (Anon.). In the last decade the growing awareness of consumers about health-related issues has not been translated into a higher consumption of fresh fruit and vegetable products. Rather, average consumption has gone down from 125 to 105 g/day whereas the recommended dietary intake is 200 g/day. This trend is also visible in other European countries. Apparently the products that are currently available no longer meet consumer demands, hence a change in product assortment is required.

From a chain-oriented perspective three major cycles of change can be distinguished (Figure 1.5). The first cycle encompasses developments in the market. A decreasing life cycle of products and rapidly changing consumer preferences lead to more impulsive behavior by consumers who have become a moving target for product developers. The second cycle refers to the technologies associated with processing and production systems. New technologies and approaches such as high-pressure cooking, hurdle technologies, minimal processing, and so on, will increase the potential to meet new consumer demands. Generally, inno-

Cycles of Change in Plant Food Production Systems

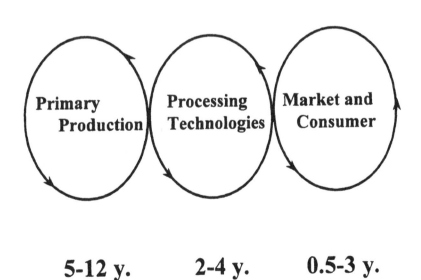

FIGURE 1.5 The three major cycles of change.

vation in technologies is slower than changes in the market situation. The third cycle deals with plant breeding and primary production and is the slowest cycle even with the use of modern biotechnology. Short-term changes in the market are impossible to follow. From a chain-oriented perspective it is of utmost importance for breeders to have a strategic view of market developments and to identify market niches where they can be strong and ahead of their competitors. Consequently, it is of great importance to establish strategic collaboration within the chain. With respect to consumption of fresh fruits and vegetables a strategic alliance between breeders/agronomists and retailers to meet changing consumer demands is vital.

QUALITY MODELING OF FOOD PRODUCTS

Development of new food products and/or varieties is based on a translation of consumer demands into product characteristics. At first selection of the desired product characteristics is done qualitatively but at a later stage these characteristics must be quantified. In an attempt to systemize this approach the system of Quality Function Deployment (QFD) has been developed (Evans and Lindsay, 1996). QFD, also called the House of Quality, is a system that puts the consumer central and is characterized by a step-by-step approach. One of the strong features of the system is that it makes considerations explicit, thus facilitating the collaboration between marketing and R&D. Also, within the agri-food sector application of QFD can assist in a more structured way of product innovation (Dekker and Linnemann, 1998).

Additionally, QFD can provide a basis for chain-oriented quality modeling. The aim is then to translate consumers' demands into specified conditions that can be used throughout the production chain. To achieve this, it is necessary to extend the QFD matrices with submodels that describe the relationship between product quality and processing and handling conditions in the chain. In the literature models are presented that describe specific quality attributes in a part of the chain. One example is the description of the Maillard reaction in liquid food products (van Boekel and Berg, 1994). Generally, chain-oriented quality modeling is based on a number of steps that start with the consumer and go back in the chain (Figure 1.6).

1. From consumer to quantitative product characteristics
2. From product characteristics to processing conditions

Quality Modeling

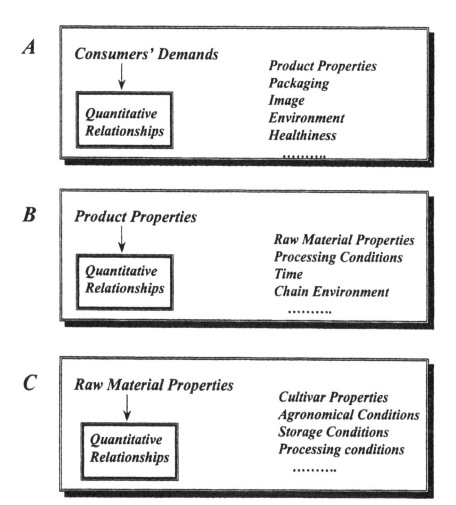

FIGURE 1.6 A quantitative approach for chain-oriented quality modeling.

3. From processing conditions to raw material properties
4. From raw material properties to breeding demands
5. From breeding demands to agronomy and storage conditions

Several attempts have been made to develop quality modeling systems at the product level. For example, Molnar, 1995 used a simplified model based on a supposed linear relationship between a quality attribute and its contribution to flavor and texture. Steenkamp and Van Trijp (1996) developed more complex multivariate correlations between consumer preferences and product properties for cheese by Hough et al. (1995) and for meat. The next step in chain-oriented quality modeling is to connect product properties to chain conditions. One example is the model that predicts the storage stability of a product with respect to spoilage and food safety as developed by Zwietering et al. (1994) and Wijtzes et al. (1995). Modeling of multivariable effects should be based on insights into the underlying mechanisms. It is to be expected that in a number of cases this kind of knowledge is not available and we will have to rely on expert systems or systems based on neural networks that use existing data to connect the demands of one actor to the specifications of the preceding actor.

LINKING CONSUMER WANTS TO TECHNOLOGIES AND RESEARCH

The previous paragraphs have emphasized that continuous innovation has become a prerequisite for companies to stay into business. This situation raises the question as to how this can happen and whether traditional innovation concepts are still valid. To answer these questions at a strategic level the DFE concept has been developed. DFE is a three-step approach that enables the establishment of linkages between changes in the market and the consequences for production systems. The approach can be used for strategic planning in supply chains of fresh fruits and vegetables.

The first step is the *Desirability* of new products (market social-economics and consumer preferences). The second assesses the *Feasibility* of the production (technological possibilities and barriers), while the third step questions *Efficacy* (organization of production chain). DFE focuses on how consumer wants can be linked to the necessary technological developments. Adequate prediction of the type of technological development needed to respond to market changes is crucial because choices should be made with regard to the type of primary product and the (combinations of) handling technologies to be used.

Recently a series of overview studies on future consumer issues have been performed for the Dutch National Agricultural Research Council

(NRLO). Meulenberg (1996) has analyzed the social-economic developments in the food market and translated them into consumer categories. Subsequently Jongen et al. (1997) and Linnemann et al. (1999) have used these categories and developed a model for translating consumer preferences and perceptions into desired technological developments. The model is based on a systems analysis, which uses the consumer as the focal point, and a stepwise approach is followed in which seven successive steps are distinguished. These steps seem to be a useful framework for an integral model of product innovation. These seven steps can be summarized as follows:

1. A thorough analysis of the social-economic developments in specified markets
2. Translation of consumer preferences and perceptions into consumer categories
3. Translation of consumer categories into product assortments
4. Grouping of product assortments in product groups according to the stages of the production chain
5. Identification of handling/processing technologies relevant for specified product groups
6. Analysis of the state of the art in relevant handling/processing technologies
7. Matching the state of the art of specified handling/processing technologies with future needs

Following this model the study showed that successfully linking R&D programs within companies to market dynamics requires a number of new technological developments. The study emphasizes the need for "dedicated" production systems that follow more closely market dynamics. A breakthrough must be realized in thinking from craft to "design for manufacture," making use of information technology and computer management systems. From a chain perspective de-coupling moments must happen as late as technologically feasible. Another outcome of the study was the conclusion that product innovation becomes more effective with respect to the success ratio when better structured. Systems such as Quality Function Deployment and Effective Consumer Response (ECR) can be valuable tools in improving chances for market success. These systems have been developed for use in the computer and automobile industry and should be developed and evaluated in order to be better suitable for use in food product innovation. The complexity of the concept of product quality in the food sector requires a specific approach and these systems should be adapted in order to be effective tools for innovation.

SUMMARY

Our knowledge about the relationships between market changes, consumer behavior, food products and handling technologies is insufficient. Nevertheless, there are promising developments, and a turning point has been reached with respect to thinking about innovation. Food production systems of the future can no longer be solely production-driven but should be primarily consumer-driven. There is a need for a chain-oriented approach to product innovation, which considers the whole food supply chain, from breeders via handling up to the consumer. One outcome should be a unified concept of product quality and acceptance throughout the production chain. Moreover, there is a need for the development of new concepts in which the various disciplinary approaches are combined into one integrated, technomanagerial approach. Technological inventions, such as in the field of biotechnology, have to be translated into products that are attractive to consumers. Changing consumer values and habits will stimulate innovation in food production technologies in order to produce suitable new products. This interdependency between consumer wants and needs on the one hand and technologies and research on the other has been recognized by many actors in the supply chain of fruits and vegetables but has not yet been implemented systematically. This interrelationship should receive more attention in the modeling of food product innovation. The stepwise approach proposed in this chapter deserves further attention in developing it into a useful approach to strategic investment in product innovation, in particular in future technologies and R&D programs.

REFERENCES

Anonymous. Zo eet Nederland. 1998. (in Dutch). 1998. Voedingscentrum, ISBN 9051770367, The Hague, The Netherlands, 216 pp.

Boekel, van M.A.J.S. and Berg, H. E. 1994. Kinetics of the early Maillard reaction during heating of milk, in *The Maillard Reaction in Chemistry, Food and Health*, P.P. Labuza et al., eds. The Royal Society of Chemistry, Cambridge, 440 pp.

Dekker, M. and Linnemann, A. R. 1998. Product development in the food industry, in *Innovation in Food Production Systems*, Jongen, W.M.F. and Meulenberg, M.T.G., eds. Wageningen Pers, The Netherlands, 67–86.

Evans, J. R., and Lindsay, W. M. 1996. *The Management and Control of Quality*, West publishing Company Minneapolis, U.S., 767 pp.

Fuller, G. W. 1994. *New Food Product Development, From Concept to Marketplace*, CRC Press, Boca Raton, U.S., 275 pp.

Hough, G., Califano, A. N., Bertola, N. C., Bevilacqua, A. E., Martinez, E., Vega, M. J., and Zaritzky, N. E. 1995. Partial least squares correlation between sensory and instrumental measurements of flavor and texture for reggianito grating cheese. *Fd Quality and Preference*, 7, 47–53.

Jongen, W. M. F., Linnemann, A. R., Meerdink, G., and Verkerk, R. 1997. In Dutch: *Consumentgestuurde Technologie-ontwikkeling; Van wenselijkheid naar haalbaarheid en doeltreffendheid bij productie van levensmiddelen*. NRLO-rapport no. 97/22, September 1997, Nationale Raad voor Landbouwkundig Onderzoek, Den Haag, 40 pp.

Jongen, W. M. F. 1995. In Dutch: *Op functionele wijze naar een gezonde toekomst*, Inaugural lecture, 27 April 1995, Agricultural University, 33 pp.

Jongen, W. M. F. 1996. Added value of agri-food products by an integrated approach, in *Agri-Food Quality*, Fenwick, G. R. et al., Eds., The Royal Society of Chemistry, Cambridge, UK, 263–272.

Juran, J. M. 1989. *Juran on leadership for Quality—An Executive Handbook*, The Free Press, New York, U.S., 368 pp.

Kooten van, O., Schouten, R. E., and Tijskens, L. M. M. 1997. Predicting shelf life of cucumbers (*Cucumis Sativus L.*) by measuring color and photosynthesis, In *Sensors for Non-destructive Testing*, Proc. Conference Orlando, Florida, 45–55. NRAES-97.

Linnemann, A. R., Meerdink, G., Meulenberg, M. T. G., and Jongen, W. M. F. 1999. Consumer-oriented technology development. *Trends in Food Science*, 9, 1–6.

Meulenberg, M. T. G. and Viaene, J. 1998. Changing food marketing systems in western countries, in *Innovation in Food Production Systems*, Jongen, W.M.F. and Meulenberg, M.T.G., Eds., Wageningen Pers, The Netherlands, 5–33.

Meulenberg, M. T. G. 1996. In Dutch: De levensmiddelenconsument van de toekomst, in, *NRLO-Rapport*, no. 96/4, November 1996, Nationale Raad voor Landbouwkundig Onderzoek, Den Haag, The Netherlands, 23 pp.

Meulenberg, M. T. G. 1997. Evolution of agicultural marketing institutions: A channel approach, in *Agricultural Marketing and Consumer Behaviour in a Changing World*, Wierenga, B., van Tilburg, A., Grunert, K., Steenkamp, J-B.E.M., and Wedel, M., Eds., Kluwer Academic Publishers, 95–108.

Molnar, P. 1995. A model for overall description of food quality. *Food Quality and Preference* 6:185–190.

Popcorn, F. 1992. *The Popcorn Report*, Harper Business, New York, U.S.

Steenkamp, J-B. E. M. and van Trijp, H. C. M. 1996. Quality guidance: A consumer-based approach to food quality improvement using least partial squares. *Eur. Rev. of Agric. Economics,* 23, 195–215.

Wijtzes, T., de Wit, J. C., Huis in 't veld, H. J., van 't Riet, K., and Zwietering, M. H. 1995. Modelling bacterial growth of Lactobacillus curvatus as a function of acidity and temperature, *Appl. Environ. Microbiol.* 61, 2533–2539.

Zwietering, M. H., Cuppers, H. G. A. M., de Wit, J.C., and van 't Riet, K. 1994. Evaluation of data transformations and validation of a model for the effect of temperature on bacterial growth, *Appl. Environ. Microbiol.* 60, 195–203.

CHAPTER 2

Quality and Breeding—
Cultivars, Genetic Engineering

P. WEHLING

INTRODUCTION

AMONG the variety of agricultural and technical factors that determine the quality of field crops, fruit and vegetables, the choice of a specific cultivar by the grower, i.e., the choice and combination of genes controlling economically important traits, may be considered the most initial step for defining quality and productivity. It is the factor that determines farmers' potential output long before any other agricultural measures are taken and even before the seed is sown in the field or greenhouse. As a consequence, the breeding of a cultivar that is adapted to specific demands of the farmer and the consumer may be considered as a preharvest factor per se, even if certain quality characters of this cultivar apply to postharvest stages.

The classical approach for breeding cultivars is to select suitable phenotypes or mutants, which are then crossed, selfed, cloned or combined with populations depending on their reproductive biology (Table 2.1). A major drawback of this approach is that for most agronomically important traits, the phenotypic variance is the base for selection, which, however, is not only composed of the genotypic variance but also comprises an environmental component as well as interactions between genotypes and environment. This tends to obscure selection progress and is one of the reasons why breeding for quantitatively or polygenically inherited traits is so tedious and time-consuming. In addition, a combina-

Table 2.1. Approaches in Plant Breeding

Classical Breeding
- Selection of phenotypes
- Intercrossing, selfing, vegetative cloning of favorable phenotypes
- Inbred line or population or clonal varieties

Marker-Assisted Selection
- Mapping of quantitative trait loci (QTL) (e.g., solids content and pH of tomato pulp, chips quality of potato, β-glucan content of barley)
- Selection of genotypes instead of phenotypes

Genetic Engineering
- Gene suppression by antisense or cosuppression
- Expression of foreign genes (constitutive or tissue-specific)
- Overexpression of native genes

tion of positively acting polygenes of one specific genetic background is difficult if not impossible because of recombinational dispersion of genes in each sexual generation. Biotechnology and molecular biology have provided exciting new tools to the plant breeder, which may help to circumvent some of the obstacles in the breeding of complex traits.

BIOTECHNOLOGY AND PLANT BREEDING

Plant breeding has been impacted during the last two decades by a number of technological developments that may enrich plant breeders' repertoire to achieve breeding progress more quickly or conveniently or even allow them to design plant traits that were impossible to create by classical breeding methods. Besides classical breeding methods, biotechnological breakthroughs like *in vitro* fusion and regeneration of plant cells, and marker-assisted selection (MAS) of monogenic traits as well as the tracing of quantitative trait loci (QTL) by use of molecular genomic markers have gained increased importance for plant breeding (Table 2.1). In tomato, e.g., QTL for quality traits such as soluble solids content, fruit mass, fruit pH, and fruit shape have been mapped (Grandillo et al., 1996). In potato, traits such as chip color, tuberization and tuber dormancy may be traced by molecular markers. For these characters, between 5 and 13 QTL have been identified. In some cases relatively high individual effects of a single QTL were found. Also, the total phenotypic variation of a given trait that could be accounted for by combined assessment of these QTLs was in the range of 50% or higher, which demonstrates the potential of molecular markers as a selection tool for tracing quantitatively inherited traits.

Limitations of MAS relate to the fact that applicability of these markers often depends heavily on the specific experimental population in which they were identified, as well as on the extent of marker polymorphism, linkage disequilibria and linkage phases among the individuals under selection. In addition, specific marker technology and know-how is required for routine marker-assisted selection programs, rendering the cost efficiency of MAS uncertain for crops that are not of major economic importance.

As a more recent achievement, transformation methods enabling the transfer of any isolated gene into virtually any important cultivated plant species have opened the toolbox of genetic engineering as a novel opportunity to the plant breeder. Plant transformation technology allows breeders to circumvent some restrictions of classical breeding methods. In particular, gene technology offers the breeder the following promises:

1. The manipulation of native and the introduction of foreign specific genes not only allow for modification of simply inherited traits like herbicide tolerance and virus and insect resistance, they also open new horizons for directed adjustment of even traits that display very complex inheritance in native systems. Thus, in some instances approaches of factorial instead of quantitative genetics may be sufficient for the breeder to cope with the improvement of specific quality- or even yield-determining characters.

2. Genetic engineering provides "added value" by transferring specific genes to cultivars that have been subject to intensive breeding efforts and, thus, are highbred already with respect to yield, uniformity, disease resistance, and so on. This aspect cannot be underestimated since any modern cultivar is defined by a complex of characters that has been assembled during years or decades of breeding work. In classical breeding new or better characters have to be introduced by crosses, giving rise to sexual recombination and, as a consequence, to dispersion of the valuable trait complexes in the progeny. In addition, undesired genes from the donor cross parent are also introduced, which have to be eliminated by successive backcrosses to the cultivar parent. It is for this reason that plant breeders tend to prevent the

introduction of "exotic," i.e., agronomically unadapted donor geno-types into their highbred elite gene pools. The addition or modifica-tion of single genes to elite lines by genetic engineering would help to alleviate these difficulties.

3. Genetic engineering speeds up the breeding process and provides bet-ter cultivars to farmers and consumers with less effort of labor and time. There is expectation that the concerted use of biotechnological methods may substantially shorten the breeding cycle for a given crop, which presently is in the range of 10–15 years or, for some woody fruit species, extends to 25 years. This time span appears quite large if one considers the need for the breeder to adapt his breeding goals to the rapidly changing requirements by growers, industry, traders and consumers.

It should be stressed, however, that biotechnological methods such as genetic engineering are not, and most probably will not become, self-sufficient in breeding better cultivars. Rather, they may provide addi-tional tools to the breeder who will still have to apply classical breeding methods, which continue to constitute the backbone of plant breeding.

The genetic engineering approach relies on (1) a detailed knowledge of the biochemical pathways that generate the quality trait, (2) the iso-lation of genes that have an impact on these pathways, and (3) the trans-fer and expression of one or several of these genes into crops in order to specifically modify the trait of interest. A number of strategies are available to genetically engineer a trait.

First, if the plant to be modified expresses a gene leading to unde-sirable characters, this gene may be shut down by introducing the same gene once more into the plant but in the opposite direction so that tran-scription of the native gene is neutralized by its antisense counterpart. Gene silencing can also be achieved by introducing a truncated version of the native gene in either direction, a phenomenon that is called co-suppression. Second, if expression and not suppression of genes is de-sired, then novel genes that are not originally owned by the plant may be introduced from any source organism and may be expressed either constitutively or specifically in a tissue or developmental stage. Finally, the plant's own genes may be expressed more abundantly by inserting additional copies of them or by combining these genes with different promotors which drive gene expression more efficiently than the gene's native promotor.

GENETIC ENGINEERING OF TRAITS AFFECTING QUALITY

Gene technology already has a strong impact on plant breeding by providing cultivars that are tolerant to herbicides, resistant to fungi, pests and viruses, or that have been rendered male sterile: Are there also implications for quality characters in crops, fruit and vegetable? From 1987 to 1997, 4,279 field trials (U.S.: 3,315 and EU: 964) at over 15,000 field sites were approved or acknowledged by the authorities. Since January 1994, 86% of all field releases in the U.S. have been approved as notifications under the simplified regulation, which has been effective since 1993. Derivatives of approximately 50 different plant species have been field-tested to date. The most frequent crops that have been subject to genetic engineering in the U.S. and Europe are summarized in Figure 2.1. Corn is the major crop being field-tested, followed by potato, tomato, soybean, rapeseed, cotton and sugarbeet. Of the field releases in EU and the U.S., 26% correspond to fruit and vegetable crops, with potato (444) and tomato (428) leading the field far ahead of melon and squash (132) and radicchio (35). Infrequent crops with five or less issued field testings include barley, broccoli, cabbage, calendula, carrot, cauliflower, cranberry, eggplant, grapevine, onion, orange, papaya, pea, pepper, plum, raspberry, strawberry, sugarcane, sweet potato, walnut, watermelon and zucchini.

These transgenic plants have been released with a wide variety of modifications. Among the field testings in the U.S., herbicide tolerance is ranked first followed by insect resistance and product quality (Fig. 2.2). Product quality comprises one fifth of the field releases and may thus be considered as a significant objective of genetic engineering. Of the commercialized transgenic crops, which amounted to 12.8 million ha (31.5 million acres) worldwide in 1997, however, quality traits have a portion of less than 1% while herbicide tolerance, insect resistance and virus resistance are the dominant traits with a portion of 54, 31, and 14%, respectively (James, 1997).

Among the fruit and vegetable crops, tomato has become the paradigm of "high-tech" plant breeding, mainly because of its relative small and simple genome, which favors gene mapping and isolation, its good *in vitro* manipulation and transformation properties, its autogamous reproductive system allowing for easy inbreeding and production of homozygous genotypes, its refractory nature with respect to crossability with wild *Lycopersicon* species, which urges the breeder to think of non-

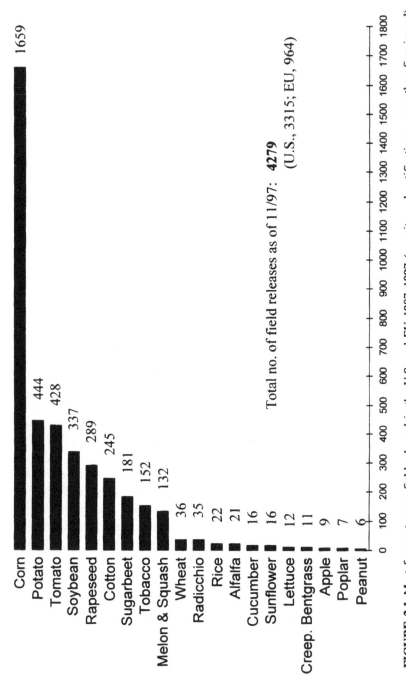

FIGURE 2.1 Most frequent crops field released in the U.S. and EU 1987–1997 (permits and notifications, more than five issued).
Sources: USDA/APHIS/BSS and RKI, 1997.

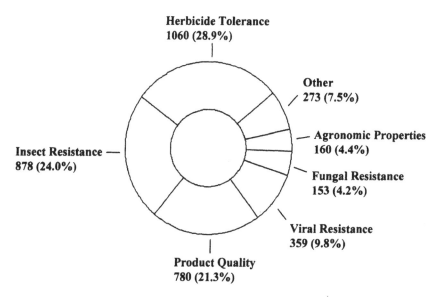

**Herbicide Tolerance
1060 (28.9%)**

**Other
273 (7.5%)**

**Agronomic Properties
160 (4.4%)**

**Insect Resistance —
878 (24.0%)**

**Fungal Resistance
153 (4.2%)**

**Viral Resistance
359 (9.8%)**

**Product Quality
780 (21.3%)**

FIGURE 2.2 Most frequent trait categories of field-released transgenic crops in the U.S. 1987–1997 (permits and notifications). *Source:* USDA/APHIS/BSS.

sexual gene transfer strategies, and because of its considerable economic importance especially in the U.S. but also in southern European countries (70 million tons produced worldwide in 1993).

The breeding goals in tomato have to be grouped by the purpose of use. Besides general goals, tomatoes have to fulfil specific demands for either processing or fresh market use (Table 2.2). Ripening characteristics, solids content, and fruit color have all been subject to genetic engineering.

Table 2.2. Breeding Goals in Tomato

General	Processing	Fresh Market
• High yield	• High solids content	• Intense, uniform fruit color
• Disease resistance	• Small and firm fruit	• Unblemished fruit
• Concentrated fruit set	• Uniform ripening	• Larger but uniform fruit
		• Long shelf life
		• Good taste

Approaches to Modification of Ripening Characteristics via Genetic Engineering

The ripening stages in the tomato fruit are referred to as immature, mature green, breaker, pink and red. Because of their tolerance to rough handling, tomatoes harvested at the mature green stage constitute the major fraction of the commercial fresh market tomato crop. Fruit harvested mature green hold the longest in storage, shipping, and on the supermarket shelf, which is of major interest for producers and industry. Ripening behavior is a compound trait and is the result of complex physiological pathways. Different approaches have been suggested to modify genes related to the ripening process. These may be grouped into efforts to reduce internal ethylene production as the physiological basis of ripening, and to delay fruit softening at the final steps of ripening. Some of the targeted genes were expected to have relevance to the major breeding goals identified by the tomato industry, i.e., viscosity, handling characteristics, soluble solids, color and taste. Moreover, from these studies, which were primarily directed toward an understanding of the ripening process in general, it was expected that approaches to target the complex trait of "ripening" would be transferable to other fruit and vegetable crops including apples, bananas, mangoes, melons, nectarines, sweet peppers, peaches, pineapples, raspberries and strawberries.

Tomato belongs to the group of climacteric fruits together with, e.g., apples, pears, melons and bananas, which develop an autocatalytic release of ethylene at the beginning of maturation. The physiological effects of the plant hormone ethylene are manifold and include seed germination, stimulation of ripening in fruits and vegetables, leaf abscission, fading in flowers, flower wilting, leaf yellowing and leaf epinasty.

Different approaches have been tested to manipulate the plant's ethylene biosynthesis (Figure 2.3). The molecular pathway of ethylene formation begins with methionine and ATP to form S-adenosylmethionine, which is converted into aminocyclopropane carboxylate (ACC) by ACC synthase. At the last step in ethylene formation, ACC oxidase is an essential enzyme in plants. Since a decrease in the rate-limiting enzymes ACC synthase or ACC oxidase should result in a concurrent decrease of ethylene production, these two enzymes have been targets in different transgenic antisense or cosuppression approaches. The transgenic tomato cultivar *FreshWorld Farms Endless Summer®* (produced by DNAP), which is a genetically engineered version of the *FreshWorld*

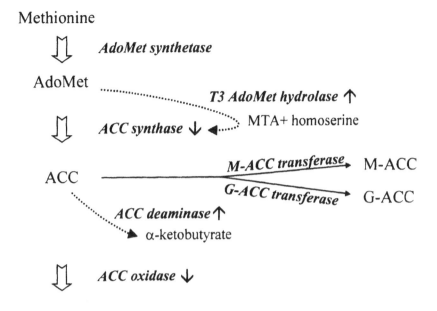

Methionine

Ethylene

FIGURE 2.3 Ethylene biosynthesis in higher plants and enzymes targeted or introduced by genetic engineering. ACC, 1-aminocyclopropane-1-carboxylic acid; Adomet, S-adenosylmethionine; M-ACC, 1-(malonylamino_cyclopropane-1-carboxylic acid; G-ACC, 1-(γ-L-glutamylamino)-cyclopropane-1-carboxylic acid; MTA, 5'-methylthio-adenosine; ↑, ↓, overexpression and suppression of targeted genes, respectively.

Farms® tomato, was put on the test market in 1995 and had been engineered by cosuppression of the ACC synthase gene (Transwitch® technology) leading to an extension of shelf life to 30–40 days postharvest. The delayed-ripening fruit ripen as usual when exogenous ethylene is applied. An alternative approach was successfully tried by introducing enzyme genes that are not originally present in tomato: In one such system (Ferro et al., 1995) exploited by AGRITOPE Inc., a gene for S-adenosylmethionine hydrolase (AdoMet hydrolase) was isolated from the bacteriophage T3 and transferred to tomato. This enzyme hydrolyzes the intermediate AdoMet by which homoserine and MTA are released. MTA in turn is a potent inhibitor of ACC synthase. Thus, the principle is again to decrease ACC activity. In another case, the enzyme gene for ACC deaminase from *Pseudomonas chloraphis* was introduced at Monsanto into tomato, which catalyzes the opening of the ACC cyclopropane ring to give α-ketobutyrate. In this case, the ethylene precursor ACC is withdrawn from the pathway, again resulting in a reduced ethylene synthesis. The genes targeted for manipulation of ethylene biosynthesis comprise AdoMet hydrolase, ACC synthase, N-ACC malonyltransferase, ACC oxidase, and ACC deaminase.

A second aspect of ripening relates directly to the softening processes in the fruit. Delay of softening has been achieved by gene suppression of enzymes that degrade the structural integrity of the cell tissue. Cellulose, hemicelluloses and pectins are cell wall components that contribute to textural characteristics. One of the pectin-modifying enzymes is polygalacturonase (PG), a pectinase, which hydrolyzes the α-1,4 linkages in the polygalacturonic acid component of cell walls. PG is synthesized specifically during ripening and is secreted into the intercellular space of pericarp cells. The function of PG during the natural ripening process is to break down the pectin of the middle lamella between the fruit cell walls, thereby causing fruit softening.

The best known case of inhibition of pectin degradation by means of genetic engineering is the antisense suppression of the PG enzyme gene, which is the effective principle in the transgenic cultivar 'FlavrSavr' by Calgene. As a different approach, cosuppression of the same enzyme gene was applied by Zeneca/Petoseed for the production of a tomato giving a higher-viscosity pulp, which was introduced into the British market as tomato paste and purée in February 1996. Suppression of pectin methylesterase (PE), which is also involved in pectin modification, and of endo-1,4-β-glucanase, an enzyme thought to be involved in fruit softening because it is known to degrade the major hemicellulosic polymer, xyloglucan, was also patented as approaches to inhibit fruit softening.

Benefits of Long-Shelf Life Tomatoes

The flavor of fresh market tomatoes is of continuous concern for consumers. Initially, the development of genetically engineered "long-shelf life" tomatoes was claimed to improve the taste of the fruit because the tomatoes would ripen on the vine for a longer time and would, thus, accumulate more flavor than traditional cultivars, which are picked at the mature green stage (Vanderpan, 1994). However, from the beginning this was a matter of controversy because tomatoes properly harvested at the mature green stage will ripen into a product indiscernible from vine-ripened fruit. Yet, it is a problem to constantly hit the appropriate, relatively shortlasting mature green ripening stage in large-scale tomato production. Fractions of fruit harvested immature may have an impact on the consumer's overall impression of taste and texture characteristics of commercial tomatoes even when they have been treated with ethylene to red ripeness. Tomato varieties with delayed ripening are, however, not expected to become a remedy for a lack of flavor because the new trait will be exploited by industry for extending shelf life after harvest rather than for longer ripening time on the vine. In addition, flavor in any fruit is a multiple-factor trait dependent on the genetic background of a cultivar and may, thus, not be expected to undergo substantial improvement by modifying a single gene. Consequently, for any of the transgenic tomatoes that exhibit reduced ethylene biosynthesis or delayed softening, improved taste may be regarded as a secondary benefit, far outweighed by the economics of improved production and processing qualities.

Approaches to Improving Processing Characteristics
via Genetic Engineering

Processing characteristics are in part related to the aspects that also apply to fresh marketability, e.g., genetic inhibition of PG expression as described above to increase the viscosity and glossy appearance of tomato paste. Solids content, which is important for processing of tomatoes, is another quality trait that has been genetically modified by different approaches. Again, pectin esterase gene suppression has been applied to increase soluble solids content by about 15%. A field release was conducted first in 1993 sponsored by the company Heinz. Invertase gene suppression (Ohyama et al., 1995; Klann et al., 1996) was approached by PetoSeeds to increase soluble solids that mainly consist of polysaccharides. Since the balance of sugars and acids has a major impact on the flavor of a fruit, attempts to modify the soluble solids are also of interest for the fresh market. Another interesting approach is to modify the expression pattern of the plant hormone cytokinin, which is thought to be involved in the generation of a metabolic sink in young reproductive tissues. As a consequence, assimilates of the plant are directed towards such a sink. The formation of a cytokinin precursor, isopentenyl AMP, is catalyzed by the enzyme isopentenyl transferase in a side chain of the carotenoid pathway (see below). To widen the source/sink relationship between the assimilating tissue and the growing tomato, fruit researchers at Calgene connected a gene for isopentenyl transferase with a promotor specific for the ovary. As a result, the sink situation in the growing fruit was pronounced, leading to significantly increased incorporation of total and soluble solids and larger sugar to acid ratios in a three-year field trial. This increase, however, was correlated with a decrease in red fruit yield and average fruit weight (Martineau et al., 1995).

Genetic Engineering: Designed Food for Health?

A key biochemical pathway in plants is the carotenoid pathway leading to the formation of two subclasses of compounds, the carotenes and xanthophylls. A side chain of this pathway leads to the formation of cytokinin catalyzed by isopentenyl transferase as mentioned above. Carotenoids provide the color pigments in a variety of fruit and vegetables such as tomatoes, parsley, oranges, pink grapefruit, spinach and red palm oil. In tomato, e.g., the red fruit color is provided by lycopene. Of the more than 600 naturally occurring carotenoids, only the carotene members α-, β- and ϵ-carotene possess vitamin A activity. These carotenes, along with γ-carotene and the carotenes lycopene and lutein, which do not convert to vitamin A, seem to offer protection against colorectal, breast, uterine and prostate cancers. Carotenes have also been related to immune response and protection of the skin against ultraviolet radiation. Additionally, they give antioxidative protection to the glutathionine Phase II detoxification enzymes in the liver and thus support elimination of pollutants and toxins from the body. The second carotenoid subclass, i.e., the xanthophylls also comprise compounds of positive biological effects such as canthaxanthin (UV protection), cryptoxanthin, zeaxanthin or astaxanthin. These compounds seem to exhibit antioxidative protection of vitamin A, vitamin E and other carotenoids. Like carotenes, the xanthophylls appear to be effective in a tissue-specific way, e.g., cryptoxanthin is ascribed a protective effect in vaginal, uterine and cervical tissues.

Carotenoids constitute an example of the growing number of plant constituents for which positive health effects are postulated. In relation to their effects or their origin, these constituents have been named "nutraceuticals," "phytochemicals," "phytonutrients," "phytofoods" or "functional foods." The definition of functional foods is still evolving. Typically it refers to foods that, by virtue of physiologically active components, combine a product's nutritional benefits with a therapeutic or health value. Such foods may be available in their natural state or may be modified, processed or genetically engineered (Anon., 1995). Despite uncertainties related to patenting, marketing and regulatory issues, nutraceuticals have gained some attention in relation to genetic engineering attempts to improve or modify the quality of agricultural and horticultural produce. Table 2.3 gives an as yet incomplete summary of known phytonutrients and attempts towards their modification via genetic engineering.

For example, by genetically manipulating the carotenoid pathway a range of different quality parameters such as contents of solids, vitamin A, antioxidative compounds, or fruit color may be altered. This has been

Table 2.3. Current Genetic Engineering of Plants for Higher Contents of Phytochemicals.

Phytochemical	Ascribed Effect
Carotenoids	↓ cancers ↑ immune system; ↑ UV protection; antioxidant; some are vitamin A precursors (α-, β- and ε-carotene)
Vitamin A	antioxidant; essential nutritive
Desaturated fatty acids	↓ LDL-choleresterol
Resveratrol	↓ cardiovascular disease
Tocopherol	antioxidant; ↑ liver detoxification enzyme activity
Phytoestrogens	↓ breast cancer
Hemicellulose	↑ binding of gallic acid
Glucosinolates	↑ liver detoxification; ↑ immune response; ↓ cancers
L-carnitine	↑ fatty acid transport across mitochondrial membranes
Epitopes of surface proteins from pathogenic microbes	↑ immune response ("edible vaccines")
Epitopes of auto-antigenes	↑ immune response ("edible vaccines")
Other applications	
Elimination of allergens	transgenic elimination of known potent food allergens (e.g., rice allergen; glutene in wheat flour)
Increase of phytase	↑ digestibility of feed and food; ↑ availability of phosphorous

↑, ↓ Increasing or decreasing effect, respectively

attempted in tomato by gene suppression or overexpression with HMG-CoA reductase or with phytoene synthase as targets, respectively (Figure 2.4). Suppression of the latter enzyme, which catalyzes the formation of the first carotene named phytoene, may lead to tomatoes that are not red but display a shade of yellow like certain peppers, which may be an

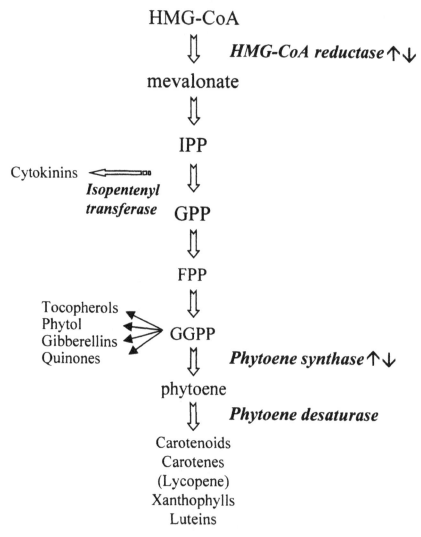

FIGURE 2.4 Major pathway of carotenoid biosynthesis in higher plants. Routes to plant hormones and to some potential phytochemicals are also indicated. FPP, farnesyl pyrophosphate; GPP, geranyl pyrophosphate; GGPP, geranylgeranyl pyrophosphate; HMG-CoA, 3-hydroxy-3-methylglutaryl-CoA; IPP, isopentenyl pyrophosphate

attractive variant to the consumer. On the other hand, overexpression of phytoene synthase may result in a deeper red color and a higher antioxidative activity of the produce (Bird et al., 1994). In rice, a phytoene synthase gene introduced from daffodil (*Narcissus pseudonarcissus*) was used to increase the level of phytoene as a β-carotene precursor in an attempt to help overcome worldwide vitamin A deficiency (Burkhardt et al., 1997). Thus, plant tissues normally devoid of carotenoid biosynthesis such as the rice endosperm can be rendered potent of vitamin A production.

There are a number of problems concerning the commercialization of phytonutrients. In contrast to "traditional" products of gene technology with novel traits introduced by novel transgenes or antisense constructs, nutraceuticals often comprise compounds which, albeit to a lesser extent, were already present in a food before the modification. As a result, eligibility for patenting may not be possible. A major drawback in the concept of phytonutrients, however, remains their ambivalent and controversial biological effects. Whereas substantial clinical data on the health impact are still lacking for most of the postulated phytonutrients, there are a number of cases where thorough reinvestigations have led to conflicting results about the physiological effects of plant components.

As a first example, linolenic acid was long considered to be a premium fatty acid in terms of nutritional benefits and attempts were made to drastically augment its contents by genetic engineering in oilseed rape. Meanwhile, further studies have come to the conclusion that well-balanced fatty acid spectra with oleic acid as the predominant fatty acid, as is found in oilseed rape of double-low quality, is of highest nutritial value. On the other hand, high-oleic sunflower oil recently was suspected to be associated with increased breast cancer rates in humans. As a second example, β-carotene as a vitamin A precursor and antioxidant was long believed to be a potent anti-cancer compound based on epidemiological data. Two clinical studies with high-risk men and women in Finland and the U.S., however, unexpectedly revealed an up to 28% higher lung cancer incidence in the β-carotene group than the placebo group (Greenwald et al., 1995). These studies, thus, place in severe doubt a beneficiary effect of β-carotenes per se and suggest other constituents or the overall effects of fruit and vegetables rich in β-carotene to have reduced cancer incidence in the older investigations. A third example relates to the trierucic content of cruciferous plants, which traditionally was associated with negative effects on heart function. In contrast, recent studies ascribe a retarding effect of erucic acid on progressive demyelination in connection with the heriditary disease ALD in male children and indicate that even high dosages of erucic acid do not seem to have adverse effects on heart health. As a consequence, dietary vegetable oils with up to 20% of pharmacologically pure erucic acid, which is provided by trierucins of cruciferous plants, are currently offered for therapeutic applications at several hundred dollars per liter. More recently, trierucins are again controversially discussed with respect to their neurological effects.

To conclude, genetic engineering of food to functional food will remain an enterprise of uncertain economic return as long as regulation does not provide market exclusivity and the health benefits (or harmlessness) of these products have not been proven by real data. Nevertheless, the expanding list of plant compounds with known or claimed health benefits along with the growing health consciousness of consumers will undoubtedly inspire genetic engineering of functional foods.

FUTURE PROSPECTS

Recent experiences have proven that besides herbicide tolerance, disease and insect resistances, quality traits are also amenable to genetic engineering. Table 2.4 gives a summary of genetically engineered quality parameters in fruit and vegetable. A selection of commercial nearterm products concerning modified quality traits in fruit and vegetables is given in Table 2.5. In the European Union, transgenic potato and radicchio with modified starch characteristics and male sterility (for use in

*Table 2.4. Genetic Engineering of Quality Parameters
in Fruits and Vegetables*

Alteration of	• Fruit softening
	• Fruit ripening
	• Storage ability
	• Total and soluble solids
	• Fruit color
	• Carotenoid contents
	• Sugar profile
	• Parthenocarpy
	• Starch branching ratio (amylose vs. amylopectin)
	• Bruising susceptibility (browning)

Table 2.5. *Commercialization of Genetically Engineered Crops in the U.S. and EU (Quality Traits Only).*

Crop	Gene/Trait	First Sales	Company	U.S. Deregulation FDA	U.S. Deregulation USDA	U.S. Deregulation EPA	EU
Rapeseed	Lauroyl-ACP thioesterase/Laurate vegetable oil	1995	Calgene	4/95	10/95	N/A	F / Review
	Phytase/Low-phytate rapeseed[a]		Mogen				NL / Review
	Barnase-barstar/Male sterility[b]	1996/1997	PGS				1996[c] / 1997
Potato	AS-starch synthase/Increased amylopectin		AVEBE				Review
Radicchio	Barnase-barstar/Male sterility	1996	Bejo Zaden				1996[c]
Tomato	AS-Polygalacturonase ('FlavrSavr')	1994	Calgene	5/94	10/92	N/A	
	CS-Polygalacturonase	1995	Zeneca/Petoseed	9/94	6/95	N/A	UK 1994[d]
	CS-ACC synthase ('Endless Summer')	1995	DNAP	10/94	1/95	N/A	
	ACC deaminase	1998	Monsanto	9/94	review	N/A	
	AdoMet hydrolase	1998	Agritope	1996	3/96	N/A	

[a]Applies to feed quality; [b]applies to seed quality and yield; [c]not (yet) allowed for food and feeding purpose; [d]cleared for food, tomato paste only; AS means antisense approach; CS means cosuppression approach.
Sources: Kridl and Shewmaker (1996); USDA/APHIS/BSS, RKI.

hybrid seed production), respectively, have been approved for commercial use. In the United States, five lines or cultivars of tomato with delayed softening and ethylene ripening characters have been approved by the FDA and USDA for commercial release.

Besides tomato and potato, a variety of other transgenic fruit and vegetables have appeared on the stage since 1995. These include broccoli, carrot, chicory, cranberry, eggplant, grapevine, pea, pepper, raspberry, strawberry and watermelon. After transgenic modifications of quality parameters had been successfully tested in tomato, the respective approaches were extended to a variety of fruit and vegetables. In 1997, field releases were conducted for apple, melon and pepper with altered product quality, for strawberry with delayed softening by transfer of a polygalacturonase inhibitor protein gene and for tomato with delayed ripening and increased solids characteristics (Table 2.6).

A promising future prospect is genetic engineering of parthenocarpic fruit, which recently was exemplified for eggplant (Rotino et al., 1997). In this case, a gene from the phytopathogenic bacteria *Pseudomonas syringae* pv. *savastanoi* encoding synthesis of the auxin precursor indolacetamide was put under the control of an ovule-specific promotor and transferred into eggplants. The transgenic plants acquired the capacity to form seedless, marketable fruits in the absence of pollination under normal growth conditions. In addition, the transgenic plants developed marketable fruits without application of exogenous plant hormones under low temperature/light conditions, which are otherwise prohibitive for fruit set in untransformed eggplants. It can be expected

Table 2.6. Field Releases for Transgenic Quality Traits in Fruit and Vegetables in 1997.

Crop	Trait	Investigator
Apple	Altered product quality (not specified)	Univ. of Berkeley
Melon	Altered fruit ripening	Agritope
Pepper	Altered fruit ripening Enhanced aroma	DNAP
Strawberry	Delayed softening (PG-inhibitor protein) Delayed ripening	DNAP
Tomato	Delayed ripening; increased solids	Zeneca

Source USDA/APHIS/BSS

that similar gene constructs will have impacts on parthenocarpic fruit production in other crops such as cucumber.

After the initial genetic engineering work in major crops, the future will certainly bring about a broadening of the range of crops that are subject to transgenic approaches. Softening inhibition, e.g., with polygalacturonase or pectin methylesterase as targets will probably be conducted in more fruits such as strawberries, grapes and possibly carrots. Similarly, ethylene biosynthesis will be altered by genetically modifying the expression of ACC synthases, deaminases, oxidases, or AdoMet hydrolases in additional climacteric fruits such as melons, bananas, pears, apples, and peaches as well as vegetables like broccoli to increase the shelf life of the produce. The fact that the effects of ethylene are quite broad opens the possibility of modifying several characters at a time by altering the plant's ethylene production. On the other hand, it means that for a given crop the transgenic manipulation may have to be fine-tuned to avoid undesirable side effects. This caution also applies to other approaches to modifying complex quality traits. For instance, initial expectations that suppression of pectin esterase would improve resistance of tomatoes to rough handling could not be verified. Even suppression of PE down to 10% of the original level did not result in reduced bruising susceptibility. The reason probably is that tissue texture is dependent on the presence of Ca^{2+} chelate complexes, which form if a sufficiently high number of adjacent carboxylic residues are present in the tissue. A high esterification state caused by reduced PE, however, results in an adverse situation (R. Carle, personal communication). The physiological complexity of most quality traits may constitute a major problem of genetic engineering, which for methodogical limits has been confined to the modification of single or few genes. Thus, limits in the extent of control of biosynthetic trafficking by genetic engineering are still significant.

Fine-tuned manipulation of biosynthetic pathways may become feasible as the control elements of these pathways are worked out in more detail. In the case of ethylene, e.g., a promising approach would be to regulate the effect of ethylene instead of turning down its abundance in the plant. As a possible way to this end the ethylene receptors of the targeted plant tissue or developmental stage may be modified to abolish their sensitivity to the plant hormone. Thus, ethylene-sensitive processes not related to the quality trait to be modified would remain untouched. An ethylene receptor and a related gene, *etr1* and *ers*, together with a number of downstream signal transduction components have already been identified in *Arabidopsis* (Fluhr and Matoo, 1996). Homologous *etr1* genes were found to be active also in other higher plants such as tomato (Zhou et al., 1996).

Besides ethylene manipulation alteration of fruit and tissue color and vitamin A content by modifiying carotenoid biosynthesis pathways will be more widely applied as well as reduction of bruising susceptibility in fruits and vegetables by suppression of polyphenol oxidase gene expression, e.g., in potato and banana. Fine-tuning of vegetable oil quality by introducing novel ACP thioesterase genes or manipulating ACP synthase, desaturase and acyltransferase genes has already proceeded considerably and brought about oilseed rape cultivars with new oil qualities such as cv. 'Laurical' (Calgene), which was sold first in 1995 and is distinguished by its capacity to synthesize lauric acid to above 40% of fatty acid contents (Kridl and Shewmaker, 1996). The detailed knowledge of the fatty acids pathway in plants allows adjustment of vegetable oil quality to the emerging demands of the consumers. As an example, vegetable oils from genetically engineered oil plants containing elevated amounts of unsaturated omega-3 fatty acids, which are essential nutrients and exert important effects on many biological processes, will certainly gain growing importance as partial substitutes for fish produce. Similarly, transgenic oilseed rape capable of synthesizing seed oil with high amounts of β-cartotene is under development for preventing vitamin A deficiency. Thus, rapeseed, soybean and flax will remain in the focus of genetic engineering approaches to tailor oil composition for specific (nutraceutical, or functional) food and nonfood applications.

A major drawback of contemporary genetic engineering techniques is the limitation imposed by the low number of genes that can be transferred at a time. This slows down the accumulation of valuable major genes in a common genetic background and precludes the manipulation of the majority of plant traits that are polygenically inherited. A promising improvement of gene transfer could come from new vectors, Binary Bacterial Artificial Chromosomes (BIBAC). BIBAC vectors allow for the transfer of at least 150 kb of foreign DNA (Hamilton et al., 1996). This opens new horizons for the transfer of large gene clusters as well as quantitative trait loci. Thus, engineering of complex traits will be made feasible as well as engineering of whole secondary product pathways or even the creation of new pathways for novel compounds.

CONCLUSIONS

Having started with herbicide tolerance and virus and insect resistances, genetic engineering of crops soon found its way to modification of quality traits in fruit and vegetables. It should be kept in mind, however, that virtually all of the present crop varieties with their impressive level of yield, resistances and quality have been bred by conventional plant breeding. From the view of the breeder, product quality is but one of a large spectrum of traits that have to be combined to culminate in a modern crop cultivar. Since most of the agronomic and quality traits are oligogenically or polygenically inherited, classical breeding methods like hybrid breeding, backcrossing or population breeding will continue to be the major basis of plant breeding in the foreseeable future. Thus, genetic engineering is generally not expected to convert a low-level cultivar into a market runner. This is illustrated, e.g., by the early 'FlavrSavr' tomato variety, which was withdrawn from the market because of its susceptibility to diseases and rough handling and unsatisfying flavor. The main task of genetic engineering in plants remains to provide an added value to an otherwise highbred crop variety. Thus, genetic engineering of crop plants will be able to unravel its full potential only if it is carefully integrated into classical plant breeding. As a consequence, since any genetically engineered plant material is withdrawn for one to several years from the continuous breeding progress that takes place in classical cultivar development, the "added value" of the transgene(s) has to be large enough to compensate for this delay and to render transgenic cultivars competitive with respect to their conventional counterparts. In the long run, genetically engineered crops will have a large potential of gaining a premium quality labeling on the market if quality traits of major public interest such as flavor, avoidance of common allergens or increased contents of nutraceutically active ingredients are consequently addressed as breeding goals. Besides monogenically inherited traits, new technical developments that allow for the transfer of a panel of selected genes at a time, together with rapidly evolving gene cloning techniques, will open the way for also targeting the major group of quantitatively inherited cultivar traits.

REFERENCES

Anonymous. 1995. Position of the American Dietetic Association: Phytochemicals and functional foods. *J. Am. Diet. Assoc.* 95: 493–496.

Bird, C. R., Grierson, D., and Schuch, W. W. 1994. Modification of carotenoid production in tomatoes using pTOM5. U.S. patent 5,304,478.

Burkhardt, P. K., Beyer, P., Wünn, J., Klöti, A., Armstrong, G. A., Schledz, M., von Lintig, J., and Potrykus, I. 1997. Transgenic rice (*Oryza sativa*) endosperm expressing daffodil (*Narcissus pseudonarcissus*) phytoene synthase accumulates phytoene, a key intermediate of provitamin A biosynthesis. *Plant J.* 11: 1071–1078.

Ferro, A. J., Bestwick, R. K., and Brown, L. R. 1995. Genetic control of ethylene biosynthesis in plants using S-adenosylmethionine hydrolase. U.S. patent 5,416,250.

Fluhr, R. and Mattoo, A. K. 1996. Ethylene—biosynthesis and perception. *Crit. Rev. Plant Sci.* 15: 479–523.

Grandillo, S., Ku, H.-M. and Tanksley, S. D. 1996. Characterization of *fs8.1*, a major QTL influencing fruit shape in tomato. *Mol. Breeding* 2: 251–260.

Greenwald, P., Clifford, C., Pilch, S., Heimendinger, J., and Kelloff, G. 1995. New directions in dietary studies in cancer: the National Cancer Institute. *Advances in Experimental Medicine and Biology* 369: 229–239.

Hamilton, C. M., Frary, A., Lewis, C., and Tanksley, S. D. 1996. Stable transfer of intact high molecular weight DNA into plant chromosomes. *Proc. Natl. Acad. Sci. USA* 93: 9975–9979.

James, C. 1997. Global status of transgenic crops in 1997. ISAAA Briefs No. 5. ISAAA: Ithaca, NY. pp. 31.

Klann, E. M., Hall, B., and Bennett, A.B. 1996. Antisense acid invertase (*TIV1*) gene alters soluble sugar composition and size in transgenic tomato fruit. *Plant Physiol.* 112: 1321–1330.

Kridl, J. C. and Shewmaker, C. K. 1996. Food for thought: Improvement of food quality and composition through genetic engineering. *Ann. New York Acad. Sci.* 792: 1–12.

Martineau, B., Summerfelt, K. R., Adams, D. F., and DeVerna, J. W. 1995. Production of high solids tomatoes through molecular modification of levels of the plant growth regulator cytokinin. *Bio/Technology* 13: 250–254.

Ohyama A., Ito, H., Sato, T., Nishimura S., Imai, S., and Hirai, M. 1995. Suppression of acid invertase activity by antisense RNA modifies the sugar composition of tomato fruit. *Plant Cell Physiol.* 36: 369–376.

Rotino, G. L., Perri, E., Zottini, M., Sommer, H., and Spena A. 1997. Genetic engineering of parthenocarpic plants. *Nature Biotechnology* 15: 1398–1401.

Vanderpan, S. 1994. A look inside a new tomato. *Sci. Food Agric. Environment* 7: 2–5.

Zhou, D., Kalaitzis, P., Mattoo, A. K., and Tucker, M. L. 1996. The mRNA for an ETR1 homologue in tomato is constitutively expressed in vegetative and reproductive tissues. *Plant Mol. Biol.* 30: 1331–1338.

CHAPTER 3

Consumer Preferences and Breeding Goals

SUSANNE PECHER
MATTHAIS VON OPPEN

INTRODUCTION

THIS chapter is about how to fulfill the requirements of consumers for quality of food products if they don't know precisely and can't express what they prefer. Consumers' preferences are crucial for the acceptance of improved varieties selected by breeders. This is especially true in low-income countries where people often depend solely on one staple food to cover their dietary needs, but holds as well for high-income countries. However, in high-income countries it is probably even more difficult for breeders to gain insight into consumers' requirements as with growing income consumer demands for food become increasingly complex. Hence consumers' preferences often remain mysterious. While breeders have skills and knowledge in the technical aspects of crop improvement, economists contribute the tools for assessing demand. It is the role of the economist to first understand and then to design ways of eliciting information on consumer preferences for quality characteristics. If this information can be translated into objectively measurable criteria, breeders can use it to select not only yield but also quality. In this interaction between plant breeders and economists it is important that both agree on certain principles that hold in the respective fields of economics and plant breeding, so that interdisciplinary exchange will be fruitful.

It is the objective of this chapter to contribute to such interdisciplinary exchange. After explaining a few basic principles of consumer de-

43

mand for quality and breeding, the concepts of hedonic price analysis and alternative approaches to assessing consumer preferences will be addressed; further, examples reported in the literature on hedonic price analyses of food products are presented; finally, conclusions are drawn.

BASIC PRINCIPLES

The Importance of Quality for Consumer Demand

According to the theory of demand, consumer choices among various consumption alternatives are guided by the principle of maximizing utility under the constraints of a limited budget. Products are chosen in such a way that the overall aggregate of consumed goods provides maximum satisfaction. To make rational choices within a limited budget, consumers rank the products (goods and services) that are offered in markets according to their contribution to their satisfaction. This ranking of products reflects the preferences of consumers. In other words: Buying is not possible without reflecting preferences.

Consumers living in industrialized market economies can usually choose among a broad spectrum of different varieties of food products. An increasingly differentiated range of processed and nonprocessed agricultural products can be found in markets, presenting a palette from convenience food over health food to exotic food. This is due, on the one hand, to an increasing depth of product processing by the agro-food industry and, on the other hand, to the overall globalization of markets. The latter makes agricultural raw products available from all over the world and independently of growing seasons.

This vast supply of food and primary agricultural products is offered at prices that have constantly fallen over the last decades in relation to consumers' income. Therefore, most consumers in high-income countries today can easily cover their dietary needs, and in most cases a point of saturation has been reached with regard to purchase of food. Hence the average household is generally not confronted with the task of fulfilling basic dietary needs, subject to its limited budget, but to choose between different qualities of food products offered, subject to health considerations and time constraints.

Beyond the basic dietary utility of food products, households in high-income countries require additional satisfaction by the consumption of food. This is, for example, the convenience of a ready-to-eat meal or the healthiness of a food product that is low in calories. Attention may be paid as well to the ethic value of agricultural raw products, produced by using techniques contributing to the protection of the environment or simply to the aesthetic value of spotless and perfect-looking fruits and vegetables. Thus, partial utilities are generated by the various qualities of products. As an example, take the different qualities of fruit juices. Fruit juice may be offered as natural juice without additives or sugar or with additional vitamins or with an exotic taste and as mixed and single fruit juice. Similar examples can be found for primary agricultural products such as the different varieties of tomatoes, asparagus and mushrooms available in markets.

For a consumer trying to maximize utility within a given budget, the absence or presence of a specific quality characteristic is often decisive for whether or not he chooses a product. Hence for producers supplying primary agricultural goods as well as processed food to markets and trying to maximize their profit, it is essential that the quality mix of the offered products is targeted to the preferences of the group of consumers envisaged.

Breeding for quality of fruit and vegetables implies that the resulting varieties will enter the market as specific product variations and hence have to meet special consumer requirements. As breeding is a long-term process, knowledge of consumer preferences is even more essential as the supplied qualities cannot be changed in the short run.

The Relevance of Quality Characteristics

Given the importance of quality, the question arises how quality can be defined and specified. In general, quality can be described as fitness for purpose (Simmonds, 1979). This understanding of quality includes two components: the objectively given characteristics of a product and the subjective perceptions of these characteristics by the consumer. The latter in turn is determined by the consumer's background of experience and information as well as by declared preferences. As reported in the literature, this leads to a variety of interpretations of quality (Brockmeier, 1993). For the scope of this chapter, the definition of quality follows the concept of economically relevant quality components (Linde, 1977; Rosen, 1974).

Products are not homogeneous but can be seen as bundles containing different quantities of characteristics. The quantitative appearance of these characteristics is objectively determined either through technical processes or by natural conditions (Lancaster, 1979). The overall sum

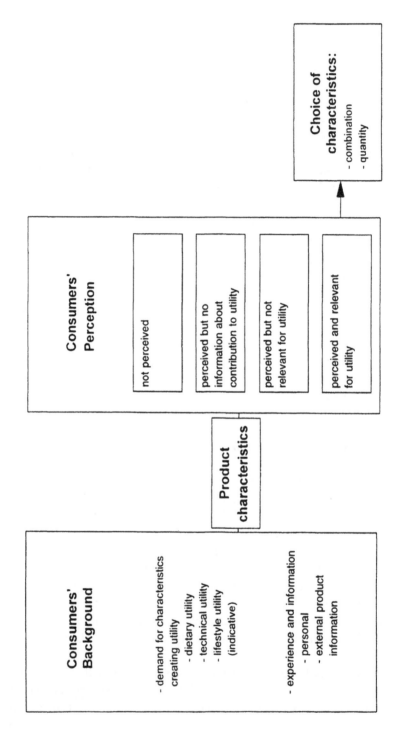

FIGURE 3.1 Relevance of quality characteristics in heterogeneous products.

of characteristics and their quantitative appearance constitutes the objective quality of a product. Differences between product variations are therefore caused by the degrees of presence or absence of characteristics. The objective quality of products is thus measurable by means of technical methods.

However, the quality of a product that is relevant for determining consumers' preferences must be defined by taking into consideration their subjective perception. Four sets of quality characteristics can be differentiated (Linde, 1977).

- quality characteristics *not perceived* by consumers
- quality characteristics that are perceived, but *not considered to have impact on utility* due to lack of information
- quality characteristics that are perceived and also considered to have impact but have *no influence on the utility* the consumer is trying to maximize, and thus not relevant
- *quality characteristics relevant* for consumers' utility maximization and thus taken into consideration in the choice process

As demonstrated in Figure 3.1, among all objectively given quality characteristics only the ones that are perceived by consumers and influence their utility are relevant for determining consumer preferences. This subset of quality characteristics can further be subdivided into evident and cryptic characteristics. Evident characteristics are directly visible, whereas cryptic characteristics are nonvisible but are perceived by consumers with the help of key characteristics such as colors (von Oppen and Jambunathan, 1978; Kroeber-Riehl, 1996). Hence the definition of quality used in this chapter encompasses all cryptic and evident quality characteristics of a product that influence consumers' utility and therefore determine preferences for a product.

CONSTRAINTS IN BREEDING FOR QUALITY

Genetic and Economic Constraints

Breeders are working in a complex system, one in which agricultural, economic and technological factors all play parts and are all in a state of more or less continuous change and interaction (Simmonds, 1979). The quality of any agricultural product is positively or negatively affected by environmental influences or by management during production, harvest, storage and processing. The objective of each breeding program is to improve the potential utility of crops. The incentives of breeding for quality, therefore, depend on the extent to which quality traits can be genetically altered, the associated costs and the market's willingness to pay for the improvement in quality. However, the baseline of each breeding program is determined by genetic and economic laws that cannot be bypassed but that can be intelligently dealt with.

Breeding targets are achieved by altering the genetic composition of crop populations with the help of selecting for visible traits. The main challenge is that the formulation of target traits as well as the process of selection is based on visible characteristics, which are determined by the invisible genetical pattern of a plant. Moreover, since most characters in living organisms are determined polygenic, the interactions of several genes with environmental effects further modify the visible expression of a genotype. This implies that selection for a specific trait cannot be done in an isolated manner but must also consider the extent to which the trait:

- is correlated with the breeding objective
- has an impact on other traits, either by negative correlation to desired traits or by positive correlation to undesired ones
- is determined by genetical influences versus the impact of environment and management

The success in breeding is determined by the returns at costs of selection for traits included in the breeding program and at opportunity costs for traits excluded. Making the right choices here is all the more critical as with an increasing number of traits for selection, time and plant materials required increase as well whereas the response to selection is likely to decrease (Simmonds, 1979). Plant breeders have to cope with a great diversity of characters determining yield and quality of a crop. The importance of each character varies in relation to the target environment and the intended use of a crop. Whereas yield is clearly defined as the crop's ability to produce biomass and to partition efficiently between productive plant parts and waste, the establishment of quality depends on the requirements of the crop users.

If a crop is used in a variety of ways, such as potato, which is used for end consumption as such, as well as for different kinds of processing, multiple quality objectives must be fullfilled. In order to effectively pursue selection for quality in a breeding program it will be necessary: (1) to set priorities between yield and quality and (2) to differentiate different objectives of quality.

Breeding is a long-term process that requires at least about 10 to 20 years' time. This implies that the objectives of breeding programs have to be based on assumptions about future developments of market demand. With regard to quality, breeders need to take a view of what will be consumers' tastes and preferences for end use as well as for processing in the coming years. However, preferences for quality traits vary due to changes in income, market supply of food products, information and technology. This often happens within a time scale that is far shorter than that of plant breeding. Therefore, a constant adjustment of objectives during the course of a breeding program is necessary. Breeders have to adjust for those characters that are no longer preferred by users and at the same time have to include new traits that markets demand.

The above considerations may be summarized by the following quotation: "Breeding is the art of throwing away" (Becker, 1993). To decide what can be thrown away, a constant flow of information about consumer preferences is required in order to identify relevant quality traits, to quantify their economic importance and to adjust breeding objectives if necessary. Furthermore, to be effective, the system of quality assessment in a breeding program must be quick, cheap and economical in the use of genetic material. These features are essential because plant breeders handle thousands of stocks each year, are often working against time, and generally have only small amounts of material available for testing (Simmonds, 1979).

Constraints on Information

Selection and breeding for food production has been farmers' preoccupation ever since their ancestors, the hunters and gatherers, settled down and began to cultivate land for crop production and animal husbandry. In such self-sufficient farms, which produced their food for home consumption, the activities of production processing and consumption as well as breeding and selection for desired quality traits were carried out by the members of one family in close interaction and breeding goals were determined in line with immediate needs. Therefore, a constant flow of information about preferred characters of crops was given.

With economic progress, commercialization and specialization removed the activities of breeding, processing and consumption from within to outside of the farm household. Breeding in specialized companies, industrial processing and urban consumption instead of food production for subsistence today offer more efficient solutions. The flow of goods is distributed from producers to consumers via markets and the corresponding counterflows of money provide price signals that allow an efficient allocation of resources. However, these flows across markets lengthen the distance between breeders and users in such a way that the required information is not necessarily passing through (Figure 3.2). Therefore, methods are required that provide the information about consumer preferences necessary for priority setting in breeding. To be applicable, these methods should be time-saving and low in costs.

APPROACHES TO IDENTIFY CONSUMER PREFERENCES FOR QUALITY CHARACTERISTICS

Informal Approaches

The approaches available for identifying consumer preferences for quality characteristics could be grouped into informal and formal approaches. Among the informal approaches one could list the individual experience of a breeder in selecting a crop. Breeders may collect informal information about the preferences in markets, households or processing enterprises by making their own observations or by exchanging knowledge with farmers, extension specialists and agricultural experts. This information collected informally might result in the breeders' personal selection index. Such informal approaches to identifying consumer preferences may be sufficient as long as yield, i.e., quantity, is the most important feature of a crop in determining gross revenue, while quality characteristics, and thus price, are of minor importance. However, once quality matters to the consumer, who is willing to pay a higher price for the satisfaction of preferences, breeders will need to apply more objective, formal approaches to assessing these consumer preferences.

Formal Approaches

Formal approaches to identifying consumer preferences are all based on the economic principles of the heterogeneity of goods and utility maximization of consumers, which were explained earlier in this chapter. These approaches can be distinguished in two subgroups: (1) approaches based on direct surveys of consumers and (2) the indirect approach of hedonic price analysis, which is based on market observations of products with different qualities and their corresponding prices. Approaches using direct consumer surveys include the conjoint analysis, factorial surveys, multidimensional scaling and consumer panels. These methods may also be combined. For a short comparative description see Table 3.1.

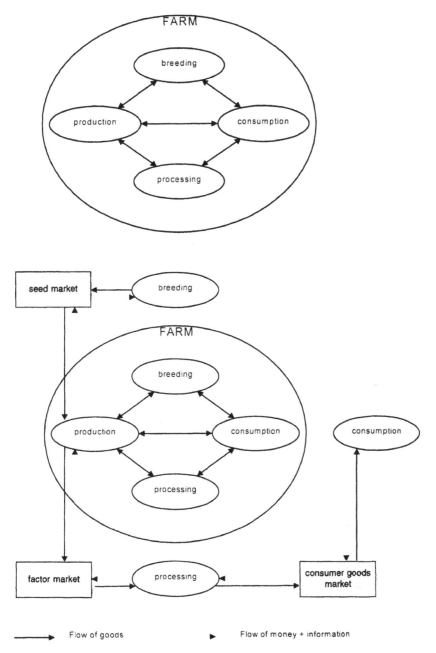

FIGURE 3.2 Flow of goods and information in subsistence oriented production systems (above) and market oriented production systems (below).

Table 3.1. Description and Evaluation of Different Approaches to Identify Consumer Preferences for Agricultural Raw Products.

Method Description and Evaluation	Classification			
	Standardized Methods		Direct	Intuitive Method
	Indirect	Consumer Panel	Formal Consumer Survey	Informal Data Collection
	Hedonic Price Analysis			
Method	Quantitative assessment of consumer prices as a function of quality traits	Testing and subsequent ranking of products by a small sample of consumers	Ranking of hypothetical products and quality characteristics by a large sample of consumers	Personal information based on experience and exchange of knowledge, informal interviews
Data obtained	Samples of products and corresponding market prices		Questionnaires with ranked data and qualitative information	Qualitative information
Analytical tools	Laboratory analysis Multiple regression	Organoleptic testing Ranking Statistical analysis	Opinion polls Ranking Conjoint, factorial analysis	Subjective evaluation
Result	Preferred quality traits in the form of measurable criteria Relevance of quality traits Index of preference for any new variety	Ranking among samples tested	Preferred quality traits in the form of descriptive terms Relevance of quality traits Index of preference for any new variety	"Personal" index of traits that are assumed to be relevant
Reliability of predicting "good quality" and representativeness of results	Very good if: Markets are competitive The product is for final consumption The product is essential	Very susceptible to: Bias due to sample selection Bias caused by panel membership	Satisfactory if: Sample selection is Representative Questionnaire fits objective	Hardly reliable not representative
Convenience for breeding	Objectively measurable criteria are directly derived Screening in early generations possible as only small amount of material is required	Time-consuming and cost-intensive Only small amount of samples can be tested Only ranking obtained	Identified traits have to be translated in measurable criteria High time requirement for preparation and analysis of questionnaire possibly	Requires long-term experience with the product Not convenient to detect recently emerged consumer preferences Not transparent

Within the standardized method hedonic price analysis has several advantages in the scope of defining priorities in breeding for quality. One of the principal advantages is that hedonic price analysis yields objectively measurable quality characters that can be directly used as criteria for selection. This is due to the fact that, unlike direct survey methods, hedonic price analysis can easily detect consumer preferences not only for evident traits but also for invisible, so-called cryptic traits. With an interview or organoleptic tests one would only find out that a certain variety is very "good for cooking" or "has a good taste." However, these approaches would not directly reveal that traits such as protein content or hardness are also significantly determining consumer preferences, as it is often the case (e.g., for sorghum in India, beans in Mexico, wheat and barley in the U.S.; Wilson, 1984; Uri et al., 1994; von Oppen and Parthasarathy Rao, 1982; Jiménez-Portugal, 1991).

Furthermore, several quality characteristics have a nonlinear relationship to consumer preferences, such that the initially positive effect may turn negative once the trait exceeds a certain amount (e.g., fat content or grain size). With the hedonic price analysis the optimum amount of such traits can be determined while this would be more complicated to detect by direct survey methods. The quality traits identified to be relevant for consumer preferences and their corresponding weights, as estimated by the hedonic price analysis, can be also combined in an economic index for the expected preference of a new variety (von Oppen and Parthasarathy Rao, 1982).

BACKGROUND OF THE HEDONIC PRICE ANALYSIS

Frederic Waugh formulated the hedonic price analysis as early as 1928 based on the observation that the different lots of tomatoes, asparagus and cucumbers in the vegetable Market in Boston, Massachusetts, showed considerable variations in price. Waugh tried to identify those quality traits that were significantly influencing daily market prices (Waugh, 1928). Since the early sixties, the hedonic price analysis has found increasing acceptance in the literature on food quality assessment. Most likely the rigor of its theoretical background and the statistical unbiasedness of results make it an attractive tool for economists (Griliches, 1990; Triplett, 1990). The possibility of deriving objectively measurable quality characteristics relevant for consumer preferences and the advantage of requiring relatively small quantities of sample material for testing are rendering this method especially interesting for breeders and producers. Today, the hedonic price analysis is a widely accepted econometric tool for the assessment of consumer preferences for quality characteristics of heterogeneous goods (Berndt, 1991; Brockmeier, 1993).

According to the theory of Rosen on market equilibrium for quality characteristics of heterogeneous goods, the hedonic price analysis is based on the following assumptions (Rosen, 1974):

- All products are perceived in markets as bundles of characteristics, which are not separable from each other.
- The transaction price of products in competitive markets can be explained by the respective quantities of different quality traits that are inherent in a product.

Hence the Hedonic price function can be written as:

$$P_{ijt} = P_{ijt}(z_{ijt1}, z_{ijt2}, \ldots, z_{ijtm})$$

P_{ijt} is the transaction price of the variation j of product i, which is available on day t in the market. z_{ijtn} is the amount of the trait n ($n = 1, 2, \ldots, m$) in one unit of the productvariation j. The explanatory variables of the function include information about the cryptic and evident quality traits in a given product that is directly or implicitly available to all market participants (Rosen, 1974; Kristensen, 1984; Triplett, 1990).

The first partial derivation of each characteristic included in the hedonic function indicates to which extent the price of a product is influenced if the amount of the respective characteristic is changed by one unit. This quantifies consumers' willingness to pay for the respective characteristic and is therefore also called the marginal implicit price. If markets are competitive, transaction prices reflect the relative preferences of consumers for products under budget constraints. Thus marginal implicit prices reflect consumers' preferences for characteristics of products and thereby provide economic weights for relevant quality traits.

These weights cannot only be used for priority setting in breeding, they also provide an economic selection index, predicting the expected preferences of consumers for new varieties. This was shown by von Oppen, who developed a preference index for sorghum in India based on marginal implicit prices and the corresponding amounts of relevant quality traits. This preference index was verified by testing 25 new sorghum varieties with a consumer panel in a double-blind test. A comparison of both rankings, the preference index and the outcome of the consumer panel, showed a highly significant correlation of 0.83 (von Oppen, 1976; von Oppen and Parthasarathy Rao, 1982).

RESULTS REPORTED ON HEDONIC PRICE ANALYSIS

A number of hedonic price analyses have been conducted to assess consumer preferences for quality characteristics in raw products that are either used for direct consumption or for processing. With products for end consumption, the hedonic price analysis has primarily focused on vegetables, fruits and meat such as tomatoes, strawberries, peaches, apples or mouton, beef and pork (Waugh, 1928; Jordan et al., 1985; Jordan et al., 1985, 1988, and 1990; Tronstad et al., 1992; Ladd, 1978; Hayenga et al., 1985; O'Connell, 1986; Williams, 1989; Wahl et al., 1995). With regard to raw products for processing, the literature mainly reports studies on wheat, barley and cotton (Ethridge and Davis, 1982; Wilson, 1984; Veeman, 1987; Larue, 1991; Bowman and Ethridge, 1992; Uri et al., 1994). Table 3.2 provides an overview of some of the reported models, showing goodness of fit as well as t-values and quality characteristics that have a significant influence on price.

Looking at these economically relevant quality characteristics shows that for products used in direct consumption, characters creating a dietary utility seem not to be as important as the evident traits determining the taste and the outer appearance of a product. The only exception here is fat content. In comparison, for the quality of raw products used for processing, cryptic quality traits influencing technical utility are clearly more significant in determining price than are evident criteria (Table 3.2).

A review of the literature on hedonic price analysis applied to derive selection criteria for breeding reveals that cryptic traits influencing dietary utility of the product are far more significant than evident traits, which have an impact on nonessential utilities such as the aesthetic appearance of the product (see Table 3.3). With the exception of groundnut and barley, most products reported are mainly used for direct consumption. Sorghum in India, beans in Mexico and rice in Thailand constitute principal staple foods. It is obvious that these studies focusing on markets of staple food products in countries with relatively low income have their own logic.

Generally, the more careful consumers are in making their choice, the more essential a good is considered. In low-income countries nutrition often depends on one staple food and in addition consumers spend a

Table 3.2. *Relevant Quality Characteristics in Hedonic Models Estimated for Different Products Used for End Consumption.*

Product	Country	Explanatory Variables and Average t-Value					adj. R²	Reference
		< 2	< 3.5	< 5	< 6.5	> 6.5		
Tomatoes	U.S.	Firmness	Color	Size			.56	Jordan et al., 1985
Lamb carcass	France			Weight Weight squared Lean color Time	Fat Fat squared Fat color	Origin Classification	.78	O'Connell, 1986
Peaches	U.S.	Color Firmness Maturity				Size Damage	.49	Jordan et al., 1987
Apples	U.S.	Origin	Trend Pesticide treatment Size month interaction	Variety Grade Season		Size Storage method	.86	Tronstad et al., 1992
Cotton	U.S.	Lot size				Grade Staple length Micronaire Micronaire squared Time	.90	Ethridge and Davis, 1982
Malting barley	U.S.	Variety		protein	plumpness plumpness sq.	Time Price feed barley	.83	Wilson, 1984
Soft wheat	U.S.		weight moisture	protein		Time Origin Dockage Foreign material Broken seeds	.88	Uri et al., 1994

considerable amount of their income on food. Consequently, it is not surprising that variations in transaction prices are significantly determined by cryptic characteristics (see Table 3.3).

In high-income countries consumers can obtain their utility-maximizing bundle of quality characteristics by consuming a variety of products. Thus, for example, if a given variety of apples does not contain the amount of vitamins required, consumers would most likely satisfy their need by buying vitamin pills instead of asking for an apple variety having a higher concentration of vitamins.

Given the vast supply of all different kinds of food products in high-income countries, the dietary utility of a specific product is less important. As expenses for food constitute only a small share of consumers' budget, dietary needs can be fulfilled in any case. Hence the average consumer in high-income countries is demanding utilities that go beyond merely the supply of nutrients. These are, for example, the healthiness of a product, influenced by the amount of nondesired ingredients, such as fat, calories or pesticides; a nice outer appearance; and a high palatability of the product. Moreover, many of these characteristics are largely influenced by management practices of production and storage rather than breeding for the right quality.

In cases where cryptic characteristics are of economic importance, and as well largely influenced by the genetic pattern of crops, appropriate quality standards and measures are necessary. These constitute the basis for the development of selection criteria in breeding. Very successful examples of this kind can be found within the group of products that are typically used for processing, such as rape, sugarbeets, wheat, barley and potatoes. In these cases the quality demanded is quite obvious as products for processing are judged by their technical performance. Subjective perceptions of consumers, which render the definition of quality difficult, is not involved in this cases.

CONCLUSIONS

When considering the established system of production and consumption of food in industrialized countries, little room is left for improvements of quality through breeding and selection for relevant characteristics. Hence, neither a direct consumer survey nor the indirect approach of the hedonic price analysis is likely to reveal quality characteristics that are sufficiently relevant to consumer preferences in order to pay off costs incurred by selecting for such traits.

Table 3.3. *Relevant Quality Characteristics in Hedonic Models Applied to Derive Selection Criteria for Breeding.*

Product	Country	Explanatory variables and average t-values					adj. R^2	Reference
		< 2	< 3.5	< 5	< 6.5	> 6.5		
Pearl millet	India	Color Glumes Broken seeds Dry volume Seed weight Swelling capacity	Protein	Moldiness			.83	von Oppen, 1978
Sorghum	India		Dry volume Swelling capacity Moldiness	Glumes Kernel weight Protein			.76	von Oppen, et al., 1982
Rice	Thailand	Chalky seeds Amylose Gel consistency Gelatinization	Shape	Color	Fragrance	Broken seeds	.89	Unnevehr, 1986
Groundnut	India	Dry matter Infestation Kkernel weight	Oil content		Origin	Broken seeds Ratio seed/pod	.71	Narasimham, et al., 1985
Beans	Mexico	Hardness Graining Shape	Color		Weight		.69	Jiménez Portugal, 1991

However, the following considerations indicate that a continuous flow of information on consumer preferences may help to increase the efficiency of breeding programs. As indicated in the basic principles described in this chapter, consumer demand for quality characteristics may change due to variations in income, preferences, technologies and product supply. Consumers in high-income countries attach relatively low importance to the nutritional value of food while other utilities such as healthiness and aesthetic value are considered to be more important. There is a general trend of demanding quality rather than quantity in food. Breeding programs should respond to these trends in order to avoid costs due to misconceptions in priority setting. However, as stated earlier, breeding for quality is not possible without concretely defining the quality traits of a product that have an impact on consumers' utility.

A changing demand for quality provoked by changes in technology has been demonstrated for cotton in the U.S. Based on the hedonic price analysis a demand function for each relevant quality characteristic was derived. It could be shown that changes in spinning technologies have a significant impact on the demand for fiber length (Bowman and Ethridge, 1992).

Changing trends of consumer preferences for fresh products can be observed in markets of many high-income countries. For example, during the last 10 years the demand for tastiness and healthiness of fresh products in Germany has increased considerably. Products that are known to be cultivated under intensive use of fertilizer and pesticides are increasingly avoided. Moreover, consumers clearly express their preferences for vegetables and fruits having a natural touch and full flavor. Jordan et al. (1985, 1987) showed that consumer preferences for various qualities of tomatoes and peaches in wholesale and retail markets are significantly determined by the characters' firmness, damage and color.

Markets are responding to these preferences by offering a broad range of different qualities of fruits and vegetables as well as an increasing share of bioproducts. Hence, for instance, it would be useful for ecologically producing farmers in Europe to have varieties that may not be top yielding but that have a solid set of resistance against major diseases, so that the cost of producing ecologically sound (i.e., chemically untreated) fruits and vegetables can be reduced.

Hedonic price analysis is an appropriate tool for early detection of changes in demand. By employing prices and supplied quantities of relevant quality traits as the information base, it directly focuses on the most essential elements that indicate trends in consumer tastes. Once the hedonic price analysis is established and validated, it is a time- and cost-efficient tool for regular follow-up surveys on consumer preferences. By attaching economic weights to relevant quality characteristics, the hedonic price analysis provides a sound basis for priority setting in breeding. Furthermore, it can be used to create economic selection indices reflecting the expected preferences of consumers for any new product or variety.

REFERENCES

Becker, H. 1993. *Pflanzenzüchtung*, Verlag Eugen Ulmer, Stuttgart, Germany.

Berndt, E. 1991. *The Practice of Econometrics. Classic and Contemporary*, Addison-Wesley Longman, Inc., Reading, MA.

Bowman, K. R. and Ethridge, D. E. 1992. Characteristic supplies and demands in a hedonic framework: U.S. market for cotton fiber attributes. *American Journal of Agricultural Economics*, 991–1002.

Brockmeier, M. 1993. *Ökonomische Analyse der Nahrungsmittelqualität*, Wissenschaftsverlag Vauk, Kiel.

Ethridge, D. E. and Davis, B. 1982. Hedonic price estimation for commodities: An application to cotton. *Western Journal of Agricultural Economics*, 7, 293–300.

Griliches, Z. 1990. Hedonic price indexes and the measurement of capital and productivity, in *Fifty Years of Economic Measurement*, Ernst B. and Triplett J., eds. University of Chicago and NBER, Chicago, pp. 185–202.

Hayenga, M. L., Grisdale, B. S., Kauffman, R., Cross, R. H. and Christian, L. L. 1985. A carcass merit pricing system for the pork industry. *American Journal of Agricultural Economics*, 315–319.

Jiménez-Portugal, A. 1991. Bohnenqualität in Mexico: Hedonische Preisanalyse. Diploma Thesis, University of Hohenheim, Stuttgart, Germany.

Jordan, J. L., Shewfelt, R. L., Prussia, S. E., and Hurst, W. C. 1985. Estimating implicit marginal prices of quality characteristics of tomatoes. *Southern Journal of Agricultural Economics*, 139–146.

Jordan, J. L., Shewfelt, R. L., and Prussia, S. E. 1987. The value of peach quality characteristics in the postharvest system. *Acta Horticulturae*, 203, 175–182.

Jordan, J. L., Prussia, S. E., and Shewfelt, R. L. 1988. A hedonic approach to estimating the value of quality characteristics of horticultural crops. *Acta Horticulturae*, 233, 376–382.

Jordan, J., Shewfelt, R. L., Garner, J. C., and Variyam, J. N. 1990. Estimating the value of internal quality aspects to consumers. *Acta Horticulturae*, 259, 139–144.

Kristensen, K. 1984. Hedonic theory, marketing research and the analysis of complex goods. *International Journal of Research in Marketing*, 1, 17–36.

Kroeber-Riel, W. 1996. *Konsumentenverhalten,* Verlag Franz Vahlen, München, pp. 266–320.

Ladd, G. W. 1978. Research on Product Characteristics: Models, Applications and Measures. Agriculture and Home Economics Experiment Station, Iowa State University of Science and Technology, Research Bulletin 584, Ames, Iowa.

Lancaster, K. 1979. *Variety, Equity and Efficiency,* Columbia University Press, New York, pp. 10–21.

Larue, B. 1991. Is wheat a homogeneous product? *Canadian Journal of Agricultural Economics,* 39, S. 103–117.

Linde, R. 1977. *Untersuchungen zur ökonomischen Theorie der Produktqualität,* J.C.B. Mohr (Paul Siebeck), Tübingen, pp. 1–30.

Narasimham, N. V., von Oppen, M., and Parthasarathy, R. P. 1985. Consumer preferences for groundnut quality. *Indian Journal of Agricultural Economics,* 40(4), 524–535.

O'Connell, J. O. 1986. Hedonic price model of the Paris carcase lamb market. *Euro. R. Agr. Eco.,* 13, S. 439–450.

Rosen, S. 1974. Hedonic prices and implicit markets: Product differentiation in pure competition. *Journal of Political Economy,* 82, 34–55.

Simmonds, N. W. 1979. *Principles of Crop Improvement.* Longman, London and New York.

Triplett, J. 1990. Hedonic methods in statistical agency environments: An intellectual biopsy, in *Fifty Years of Economic Measurements,* Ernst, B. and Triplett, J., eds., University of Chicago and NBER, Chicago, pp. 207–233.

Tronstad, R., Huthoefer, L. S., and Monke, E. 1992. Market windows and hedonic price analyses: An application to the apple industry. *Journal of Agricultural and Resource Economics,* 17(2), 314–315.

Unnever, L. J. 1986. Consumer demand for rice grain quality and returns to research for quality improvement in Southeast Asia. *American Journal of Agricultural Economics,* 68(3), 634–642.

Uri, N. D., Hyberg, B., Mercier, S., and Lyford, C. 1994. The market valuation of the FGIS grain quality characteristics. *Applied Economics,* 26, 701–712.

Veeman, M. M. 1987. Hedonic price functions for wheat in the world market: Implications for Canadian wheat export strategy. *Canadian Journal of Agricultural Economics,* 35, S. 535–552.

von Oppen, M. and Jambunathan, R. 1978. Consumer Preferences for Cryptic and Evident Quality Characters of Sorghum and Millet. Paper presented at the Diamond Jubilee Scientific Session of the National Institute of Nutrition, Oct. 1978, National Institute of Nutrition, Hyderabad, India.

von Oppen, M. and Parthasaraty, Rao P. 1982. A market-derived selection index for consumer preferences of evident and cryptic quality characteristics of sorghum, in *Proceedings of the International Symposium on Sorghum Grain Quality,* 28–31 Oct. 1981, Patancheru, A. P. India, ICRISAT, pp. 354–364.

von Oppen, M. 1974. Evaluation of relevant economic characteristics of legumes and cereals in the semi-arid tropics. Mimeographed. ICRISAT, *Econonics Program Report,* Patancheru, A. P., India.

von Oppen, M. 1976. Consumer Preferences for Evident Quality Characters of Sorghum. Paper presented at the Symposium on *Production, Processing and Utilization of Maize, Sorghum, and Millets,* Central Food Technological Research Institute, Mysore, India.

von Oppen, M. 1978. A preference index of food grains to facilitate selection for consumer acceptance. ICRISAT *Economics Program Discussion Paper,* No. 7, Patancheru, A. P., India.

Wahl, T. I., Shi, H., and Mittelhammer, R. C. 1995. A hedonic price analysis of quality characteristics of Japanese Wagyu beef. *Agribusiness,* 11(1), 35–44.

Waugh, F. V. 1928. Quality factors influencing vegetable prices. *Journal of Farm Economics,* 10, 185–196.

Williams, C. H. 1989. The choice of functional form for hedonic price functions: An application of the box-cox transformations to cattle prices in Queensland. *Discussion Paper in Economics,* University of Queensland, 15, S. 1–19.

Wilson, W. W. 1984. Hedonic prices in the malting barley market. *Western Journal of Agricultural Economics,* 29–40.

COMMON GROUND

- Commercially distributed fresh fruits and vegetables are not meeting expectations of consumers.
- Poor quality of fruits and vegetables is a complex problem and cannot be attributed to a single factor.
- A limiting factor in quality performance during marketing and distribution is the genetic potential of the harvested cultivar.

DIVERGENCE

- The focal point for integrating research to improve quality of fresh fruits and vegetables.
- The relative importance of management decisions and genetic potential in quality of fresh fruits and vegetables as delivered to the consumer.
- The ability of conventional or molecular approaches to increase the economic value of fruit and vegetable crops.

FUTURE DEVELOPMENTS

- A conceptual framework to integrate research on quality of fresh fruits and vegetables.
- Studies linking preharvest factors to postharvest quality of fresh fruits and vegetables.
- Quantitative models that link economic value to consumer preference and to cultivar selection.

FRUITS AND VEGETABLES

CONTEXT

- Genetic factors contribute to variation in quality of a fruit or vegetable.
- Cultural and environmental factors in the field limit quality potential during marketing and distribution.
- Handling and storage techniques have been designed to minimize quality losses between harvest and purchase.

OBJECTIVES

- To describe factors before and after harvest that affect the storage quality of apples.
- To present potential chemical markers for environmental factors that affect the postharvest quality of vegetables.
- To describe the interaction of physiological processes and environmental conditions during storage of fresh vegetables as it affects product quality.

CHAPTER 4

Effects on the Quality of Stored Apple Fruit

DAVID S. JOHNSON
MARTIN S. RIDOUT

MUCH of this chapter is concerned with apple storage quality since this is our predominant area of interest. In addition, our experience has been limited mainly to apple cultivars grown in the temperate climate of the U.K. However, the principles of interactions between pre-harvest growing conditions and the postharvest storage environment are similar for other applications, and need to be understood, so that the challenge of providing quality to the consumer can be met consistently.

Variation in the quality of different consignments of apples of the same cultivar from commercial storage is of major concern to producers, marketing organizations and consumers. In the U.K., much of the concern is related to consignments of fruit that have an insufficiently green background color in those cultivars where this is considered to be an indicator of freshness such as Cox's Orange Pippin (Cox) and, more importantly, to the likely texture, taste and flavor of the fruit. In the U.K. and other parts of Northern Europe, there is a strong preference for firm, juicy apples. Textural quality is generally more important than aromatic properties. While consumers in different countries may have differing preferences for the various quality attributes, there is a universal need to produce fruit that is free from physiological disorders and pathological decay.

Methods for preserving fresh fruit and vegetables are well developed and rely principally on reduction of the respiration rate of the product by lowering the temperature and by restricting oxygen availability and/or elevating carbon dioxide concentration in the storage environment. Although these technologies are not new, continual improvements in refrigeration and controlled atmosphere (CA) technology have resulted in increased quality of fruits and vegetables and further extension of storage life (Hardenburg et al., 1986). Other technologies, such as ethylene removal, offer the prospect of even greater retention of quality in store (Knee et al., 1985). Improvements in postharvest storage is a "target" for the genetic manipulation of apple and other crops. Particular attention has been paid to slowing the rate of ripening of climacteric fruits through control of ethylene synthesis (Tucker, 1994).

With the high investment costs associated with modern storage techniques, it is imperative that maximum revenues are realized on fruit from store. This requires freedom from physiological disorders up to the point of consumption and an eating quality of all consignments that is above the minimum required by consumers. Products such as apples are intrinsically variable in quality and storage potential at harvest. It is particularly important therefore that storage conditions and duration of storage are compatible with achievement of the required quality in all consignments. Substandard quality in the market place during the normal period of availability for a given cultivar can have adverse effects on subsequent sendings. This could result ultimately in total rejection of any further consignments, regardless of their quality.

The outturn quality of individual consignments of apples and pears, and presumably also other types of fruit, is dependent on their intrinsic potential to store and on their response to the imposed postharvest environment. It is important therefore that research is carried out that recognizes the continuum between pre- and postharvest physiology. It may then be possible to assess the storage potential of consignments at, or in advance of, harvest and to decide on the most appropriate conditions and duration of storage. Such research may also suggest ways of improving the quality of fruit stored in a particular regime.

CULTIVAR DIFFERENCES IN APPLE QUALITY

There are innate differences in the quality of apples of different cultivars that characterize their eating quality and their potential for long-term storage. Cultivars also differ in their propensity to commence autocatalytic ethylene production, for example, Gloster69 is particularly reluctant. This heritable trait can be used in conventional breeding to produce cultivars with improved storage quality (Stow et al., 1993). Fruit measurements taken at commercial harvest for long-term storage

(Table 4.1) can be related to the textural and taste characteristics of particular cultivars. Gala is typically crisp and sweet, Red Pippin (Fiesta) crisp and acidic and Jonagold softer in texture and balanced with regard to sugar and acidity. As fruits ripen slowly during storage, production of aroma compounds adds further to the recognition of any particular cultivar. In addition to such fruit characteristics, apples of different cultivars vary in their potential to store and in their response to methods to prolong storage life. These cultivar-storage environment interactions are particularly important since they determine tolerances to the stresses that are imposed by low storage temperatures and by low oxygen and elevated carbon dioxide concentrations in the storage environment (Johnson, 1994a). Differences in the fruit structure, physiology and biochemistry of apple cultivars have promoted the need for research to suit storage conditions to each cultivar. Although clonal differences in the

Table 4.1. Harvest Quality of Apples from Commercial Orchards in the U.K. in 1996.

	Red Pippin	Gala	Jonagold
Nos. Orchards	13	12	12
Harvest date	12 September	24–25 September	1–2 October
Firmness (N)			
mean	79.9	80.3	72.4
range	18.7	14.9	9.3
standard deviation	6.32	5.04	2.64
Acidity (mg g^{-1})			
mean	6.3	3.8	6.3
range	1.6	1.0	2.9
standard deviation	0.43	0.29	0.77
Total sugars (%)			
mean	9.5	11.7	11.1
range	2.3	1.9	2.7
standard deviation	0.56	0.68	0.79
Reducing sugars (%)			
mean	5.0	7.8	8.3
range	0.9	0.7	2.1
standard deviation	0.30	0.21	0.59

storage and eating quality of apples have been demonstrated (Smith and Stow, 1985), storage conditions have generally not been optimized for specific clones.

VARIABILITY IN QUALITY PARAMETERS AT HARVEST

At the time of commercial harvest, apples of the same cultivar from different orchards vary in visual quality such as size and red color. Also, they vary in terms of eating quality such as concentrations of sugars and acids, and fruit texture. Quality and degree of ripeness of apples on a single tree is also variable and is influenced particularly by shade and position within the canopy (Jackson et al., 1971). Ripening of individual Cox apples occurs typically over a period of three weeks (Johnson, 1995). Variation in the quality and physiological stage of individual fruits within a tree is compounded by overall quality differences between orchards. The average, range, and standard deviation for measurements of firmness, acidity and sugars in Red Pippin, Gala and Jonagold apples from several orchards are presented in Table 4.1. Data presented in Figure 4.1 as box plots indicate the extent of orchard variability in the concentration of soluble solids and titratable acid, and in the firmness of Cox apples before (pick 1), during (picks 2 and 3) and after (pick 4) commercial harvest for long-term CA storage in 1995. Data are for Cox apples from 32 commercial orchards throughout the major production areas of the U.K. In these box plots, the box indicates the upper and lower quartiles of the distribution of the values, and the line within the box indicates the median. The vertical lines extending above and below the box indicate the 90% and 10% points of the distribution. Clearly, some account needs to be taken of the variability in these quality criteria and of the marked change in the levels of firmness, soluble solids and acidity as harvest is delayed. Loss of firmness and acidity occurs during storage and the extent of this loss relates to the duration of storage and the storage conditions that are imposed. An important approach to the achievement of the desired quality in fruit ex-store is to relate harvest quality parameters to those of stored fruit. In several countries, notably the U.S., South Africa and the U.K., programs are operated to identify optimum harvesting dates for different orchards/growing regions to achieve the required quality ex-store. Identification of suitable harvesting periods remains an important area of research in Europe (De Jager et al., 1996), the U.S. (Silsby, 1993) and elsewhere. Relationships between fruit parameters at harvest and ex-store quality will be consid-

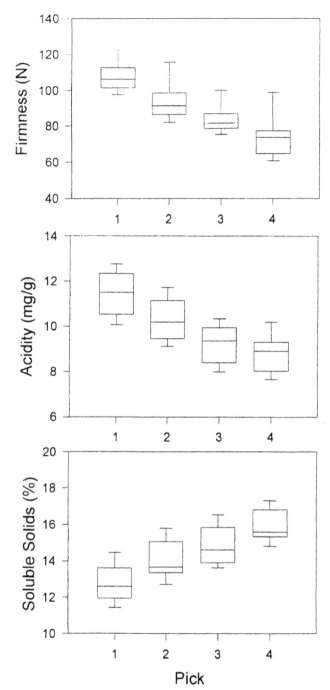

FIGURE 4.1 Box plots describing the variation in firmness, acidity and soluble solids concentration in Cox's Orange Pippin apples from 32 commercial orchards in 1995. Fruit was harvested on 31 August (pick 1) and on 11 (pick 2), 21 (pick 3) and 29 (pick 4) September.

71

ered more fully later. Briefly, to achieve consistent quality from store, full account must be taken of the considerable variation in quality attributes and the physiological state of the fruit at the time of harvest.

VARIABILITY IN QUALITY PARAMETERS AFTER STORAGE

Apples that are destined for long-term CA storage are normally harvested prior to or at the onset of the exponential rise in the rate of respiration (climacteric). This coincides with a rapid increase in ethylene production and the transition from fruit development to fruit ripening. Under the influence of low storage temperatures, restricted oxygen and/or elevated carbon dioxide concentration, the rate of respiration of the fruit is reduced and the rate of ripening is retarded (Fidler, 1973). During storage, apples and many other types of fruit gradually soften, become more yellow (primarily due to chlorophyll loss) and decrease in acid content. The extent of development of the aromatic flavor is highly dependent on cultivar, harvest maturity and duration and regime of storage.

In addition to changes in fruit chemical composition and structure that affect quality, physiological disorders may develop in the fruit during storage. Certain disorders arise through inherent deficiencies in the fruit at the time of harvest. For example, bitter pit is associated primarily with deficiency of calcium; this results in dysfunction of cell membranes and eventual death of groups of cells and the production of characteristic "corky" lesions in the fruit flesh (Perring, 1986). Other disorders develop progressively as storage is extended, e.g., senescent breakdown and superficial scald are affected principally by harvest date, storage conditions and time in store (Wilkinson and Fidler, 1973). Preharvest conditions during development often have a major influence on susceptibility of fruit to these types of disorders (Sharples, 1973). A further category of disorders includes those that are induced by the storage conditions themselves (Lidster et al., 1990). These can occur when recommended storage conditions are applied, e.g., low temperature breakdown (LTB) in Bramley's Seedling (Bramley) apples and late storage corking in Cox apples, but result more usually from failure to control carbon dioxide and oxygen levels.

Figure 4.2 indicates the extent of variability in the quality of Cox apples harvested from 24 orchards over a six-year period. Within each year all orchards were harvested on the same date, and in all years fruit was

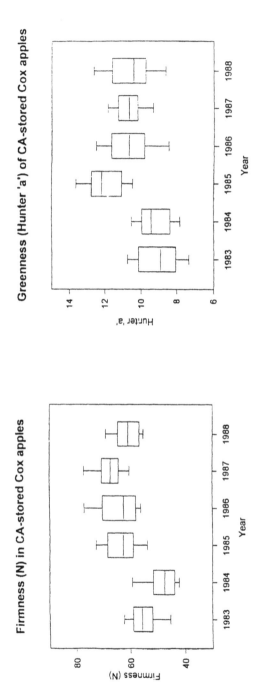

FIGURE 4.2 Boxplots indicating orchard-to-orchard and year-to-year variation in some quality attributes of Cox apples from 24 commercial orchards. Reproduced with the kind permission of *The Journal of Horticultural Science & Biotechnology* (Johnson and Ridout, 1998).

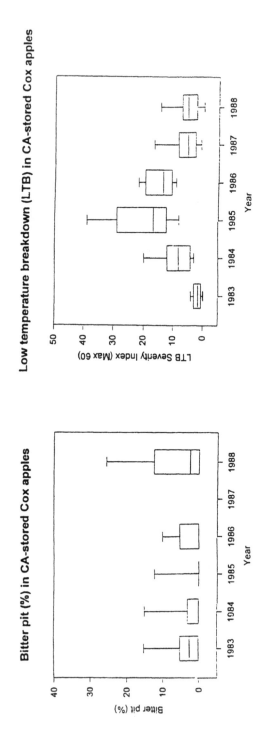

FIGURE 4.2 (*Continued*)

74

stored in the same conditions (1.2% O_2, < 1% CO_2 at 3.5°C) for the same period of time (35 weeks). The box plots provide a good illustration of the degree of variability in the primary market quality attributes, firmness and greenness, and in the major physiological disorders, bitter pit and LTB (data for LTB relate to fruit kept at 2.5°C to induce the disorder).

For each apple cultivar, the attributes associated with sensory quality, such as firmness and the balance of sugars and acid, vary according to orchard and storage conditions imposed. Table 4.2 illustrates the ex-

Table 4.2. Quality of Red Pippin, Gala and Jonagold from Commercial Orchards after CA Storage for 36, 27 and 39 Weeks, Respectively (Firmness Data for Jonagold Relate to Fruit Stored for 30 Weeks).

	Red Pippin		Gala		Jonagold	
Nos. Orchards	13		12		12	
CA conditions ($\%CO_2$:$\%O_2$)	0/1.2	3/2	0/1	5/1	0/1.2	4/1.2
Storage temperature (°C)	3.5	3.3	1.3	1.3	1.3	1.3
Firmness (N)						
mean	68.6	69.2	69.2	76.9	61.7	74.3
range	10.8	15.7	19.9	20.4	18.0	6.17
standard deviation	3.24	4.20	6.36	6.38	5.63	2.22
Acidity (mg g^{-1})						
mean	4.2	4.3	3.1	3.2	4.2	4.1
range	1.0	1.4	1.4	0.8	2.4	2.3
standard deviation	0.26	0.46	0.41	0.27	0.58	0.66
Total sugars (%)						
mean	9.6	9.6	12.3	11.3	10.7	10.8
range	1.7	1.7	3.1	3.0	2.4	3.2
standard deviation	0.53	0.50	0.95	0.98	0.76	0.96
Reducing sugars (%)						
mean	7.4	7.6	9.9	9.2	9.7	9.8
range	0.8	1.4	1.5	2.1	1.9	1.9
standard deviation	0.26	0.42	0.48	0.68	0.56	0.69

tent to which quality attributes in Red Pippin, Gala and Jonagold apples vary when stored under different CA regimes. In these examples, different CA conditions had little effect on the average concentration of sugars and acids although improvements in the firmness of Gala and Jonagold were achieved by maintaining a higher carbon dioxide concentration in the presence of 1% oxygen. In Gala apples, the variability in firmness of the fruit was similar in both CA regimes tested, whereas in Jonagold apples the presence of carbon dioxide increased average firmness and decreased variability. This result indicates an interaction between consignments and the storage environment.

AN APPROACH TO ACHIEVING CONSISTENT QUALITY FROM STORE

Because of the variability in quality that exists in fruit at the time of harvest and ex-store, it may seem an insurmountable task to provide fruit of consistently high quality to the consumer during the marketing period for any particular cultivar. With current storage technology, some change in the quality characteristics of the fruit during storage is inevitable, but it is possible to define a storage period and storage conditions that will generally provide the quality expected of any given cultivar. Postharvest research has led to a continual improvement in storage technology and to the provision of recommendations that are expected to induce no injury to the fruit and to provide fruit that are free of physiological disorders and are of acceptable eating quality. It is particularly important to include fruit from a range of orchards at an early stage in storage experiments and to carry out trials over several years so that storage behavior can be assessed adequately.

Usually, some knowledge of the composition and physiological state of the fruit at harvest is critical to achieve acceptable eating quality and minimal development of physiological disorders in the stored fruit. Once the most appropriate storage conditions and termination dates for a particular cultivar have been established, the extent of variability in storage quality needs to be quantified. If consignments are identified as having unsatisfactory quality, further work is needed to investigate the cause of the problem and to determine how to maximize the storage potential of the fruit. This approach requires a pooling of knowledge from physiologists involved in plant development and those concerned with postharvest physiology. An integration of skills and a realization that quality out of store begins with fruit development is necessary. This appreciation is consistent with the theme of this book.

It is well documented that the stage of maturity at harvest has a critical effect on storage life and quality. This is not surprising since the rapid rise in ethylene production associated with the climacteric is considered to trigger softening in apple. In other fruits, such as tomatoes and avocados, enzymes involved in ripening are induced at the onset of the climacteric (Christoffersen et al., 1982; Grierson et al., 1985). Considerable research has been undertaken to relate quality characteristics of fruit after storage to direct indicators of maturity (e.g., proximity to the climacteric and associated rise in ethylene production). Establishing relationships between quality attributes of fruit after storage and "maturity" indicators at harvest provides an opportunity to create harvest maturity standards as a means of providing more consistent quality. For example, Knee and Farman (1989) developed regression equations to predict firmness of Cox apples after CA storage on the basis of harvest firmness and ethylene concentration in the fruit. One of the complications in the development of picking date criteria is that the various quality attributes of stored fruit have different optima. Early harvested fruit will tend to be firmer, more acid, less sweet and less aromatic than fruit that are harvested later. The data in Table 4.3 show the effect of harvest date on the average quality of Cox from 30 orchards stored under the same CA conditions. Although fruit from the first harvest were high in firmness, the sensory scores were low, particularly for aroma. It is difficult to establish a harvest date that is optimal in all respects. In this situation, there has to be a decision on the priority of the various quality attributes and a compromise must be made to achieve the best all-round quality.

To achieve consistent quality in fruit from store, it is not sufficient to consider harvest maturity/quality criteria. For some quality attributes such as firmness, much of the variation ex-store can be accounted for by measurements carried out at harvest (Knee and Farman, 1989; Silsby, 1993). However, for some other quality attributes and for certain physiological disorders that develop during storage, it is necessary to consider the chemical composition of the fruit during development. Many authors have reported the importance of mineral nutrition on the susceptibility of apples to storage disorders (Sharples, 1980; Bramlage et al., 1980). Possibly the most researched disorder is bitter pit, which relates predominantly to the inadequate supply of calcium to the developing fruit (Perring, 1986). Data are available that link the susceptibility

Table 4.3. *The Effect of Harvest Date in 1994 on the Storage Quality of Cox's Orange Pippin Apples Stored in 1.2% O_2 <1% CO_2) at 3.5°C until Early January. Figures in Parentheses Refer to Fruit Kept in Air for a Further 7 Days at 18°C to Simulate Marketing. (Data Are Means for 30 Orchards Except for the Final Pick, where Fruit Was Unavailable from Two Orchards.)*

	Harvest Date			
	1 September	12 September	22 September	3 October
% Rotting	0 (0.7)	0.5 (1.3)	0.7 (1.2)	18.1 (35.6)
Greenness Color score:				
1-green 4-yellow	1.2 (2.0)	1.7 (2.2)	2.1 (2.4)	2.8 (3.4)
Hunter "a"	−13.7 (−11.2)	−12.6 (−10.8)	−11.0 (−10.0)	— (—)
Firmness	72.0 (66.3)	63.7 (58.8)	59.4 (56.0)	54.3 (52.2)
Taste:				
Sugar/acid (max 5)	(2.5)	(4.7)	(4.9)	(3.2)
Texture (max 5)	(2.7)	(4.3)	(3.3)	(2.3)
Aroma (max 10)	(2.3)	(5.4)	(6.4)	(6.3)

Hunter "a"—the more negative the figure, the greener the fruit. Color score 3 is too yellow for premium quality. Firmness measured as Newtons (N) using an automatic penetrometer fitted with an 11 mm diameter probe. −red color had developed to an extent that precluded measurement of background color.

of fruit to specific disorders with the concentration of mineral elements in the fruit at particular stages during development. In many cases, it has been possible to produce standards for mineral composition of fruit that allow fruit growers to rank orchards in terms of their storage potential and to decide on the most appropriate storage conditions and duration of storage for each consignment (Johnson, 1989).

While mineral composition is important in relation to certain physiological disorders, for others, e.g., superficial scald, correlations with the concentration of mineral elements in the fruit are poor (Johnson and Ridout, 1993). For this disorder, climatic conditions during fruit development provide a good indication of the susceptibility of different consignments of apple and it has been possible to develop prediction models based on the behavior of fruit from many orchards monitored over several years. Similarly, it has been possible to develop a regression model based on climatic variables to predict LTB in Bramley apples (Johnson et al., 1989). It is a challenge for future research to understand the effects of climatic changes on fruit physiology and biochemistry. Recent research on scald provides a good example of this approach, for which changes in antioxidant activity and extent of alpha-farnesene oxidation in the peel of apples are used to determine the effects of preharvest factors (Barden and Bramlage, 1994). This approach will provide opportunities for more accurate prediction methods based on analyses of fruit, and for the modification of the development of the fruit to confer resistance to disorders and to improve quality in store. The development of models to predict storage quality based on fruit composition and climatic variables represents a major step towards achieving consistent quality from store (Johnson and Ridout, 1998), but further integration of the factors that influence quality, and its loss in store, is needed before all aspects of fruit quality are predictable.

RESEARCH TO MINIMIZE OCCURRENCE
OF BITTER PIT IN COX APPLES

It is useful to provide an example of how this approach has minimized the problem of bitter pit in stored Cox apples. Storage experiments with fruit from different orchards over several years established that CA conditions affected the potential of fruit to develop bitter pit (Johnson, 1989). Extensive survey experiments provided the data to develop relationships between the incidence of bitter pit and the mineral composition of fruit and to establish fruit analysis standards appropriate for different methods of storage (Sharples, 1980; Johnson, 1989). Analytical services were provided to growers, and advice was given on prediction of bitter pit risk and management of consignments (Waller, 1980). Research was carried out to understand the influence of orchard factors on the mineral composition of fruit, and of calcium in particular, that led to changes in orchard management practices to enhance calcium uptake in the developing fruit.

Application of orchard sprays containing calcium is now routine in U.K. orchards, and in most other apple-growing regions of the world. Improved control of bitter pit has been achieved by changes in storage technology, particularly the adoption of automated "ultra low oxygen" (ULO) storage, and by prediction models that integrate climatic and nutritional data that accounted for up to 67% of the variance in bitter pit in air-stored Cox apples and 39% for similar samples stored in ULO conditions (Johnson and Ridout, 1998). As a result of an integrated approach, bitter pit is now seldom a commercial problem in CA-stored Cox apples in the U.K.

RESEARCH TO IMPROVE THE FIRMNESS OF COX APPLES

Although changes in storage technology and the adoption of ULO have improved the textural quality of Cox apples after storage (Sharples, 1982), variation in the firmness of consignments of apples removed from store remains a significant problem. Regression models that predict the firmness of Cox apples after CA storage, based predominantly on harvest firmness, have been developed from a three-year program of work where storage behavior of apples from up to 32 orchards harvested over a four-week period has been evaluated.

Figure 4.3 shows the relationship between harvest and ex-store firmness for fruit harvested in 1994. Fitting a regression model based on harvest firmness only to data over a three-year period (1994–96) accounted for 56% of the variation in ex-store firmness. Models that predict firmness have also been developed from larger data sets, which include chemical and physical measurements made on developing fruit in 24 orchards over a six-year period. A number of models have been developed to predict the firmness of fruit from store. The inclusion of nutritional and climatic variables with harvest firmness accounted for 76% of the variance in the firmness values of CA-stored Cox apples (Johnson and Ridout, 1998). This research has prompted further investigation of the factors

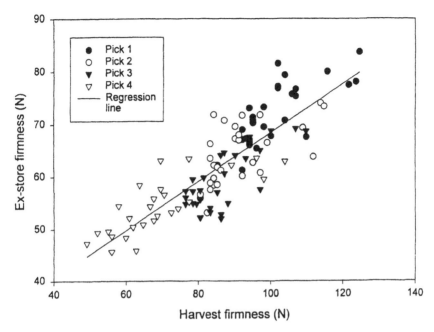

FIGURE 4.3 Relationship between the firmness of Cox's Orange Pippin at harvest in 1994 and after storage in 1.2% O_2, <1% CO_2 at 3.5°C until January 9th, 1995. Fruit was harvested on 1 (pick 1), 12 (pick 2) and 22 (pick 3) September, and on 3 October (pick 4)

that influence the mechanical properties of fruit during development with a view to improving firmness at harvest and ex-store. It is clear from recent work that under U.K. conditions higher yields per tree can result in adverse effects on fruit texture, and significant improvement in firmness of fruits has been achieved through thinning of the crop at specific stages of development (Johnson, 1994b). The increase in firmness coincided with increased dry matter and alcohol-insoluble matter. This suggests that crop load is a factor that determines the amount of photosynthate that is available for incorporation into the cell walls of the developing fruit. Cultural practices that are likely to encourage water uptake and depress the accumulation of dry matter in the developing fruit, such as excessive irrigation or extensive use of herbicides in orchards, should be avoided (Richardson, 1986). This is a good example of how a postharvest quality problem is being tackled from a consideration of all aspects of apple development, maturation and ripening. In this case, consistent textural quality in Cox apples is only likely to be achieved by an integrated approach involving the pooling of expertise and knowledge from several disciplines.

REQUIREMENTS FOR FUTURE RESEARCH

It should be clear from the preceding sections that the innate variation in a crop of apples harvested in each year represents a considerable challenge to the provision of a consistent, high-quality product to the consumer. The ultimate aim is to understand the effects of all the factors that impact upon the fruit during development and that influence quality after harvest and during storage. The ultimate aim must be a prediction of all the attributes that constitute quality, assuming that there is agreement on the definition of quality. The changes in consumer requirements that inevitably occur will demand a continued effort by researchers to understand the underlying mechanisms and processes that determine quality. There is a need to understand the basis of the variation in the quality of individual apples within a population on a tree, in addition to the average quality of the population. Improved prediction of quality and disorders will undoubtedly require considerable resources to ensure that all relevant explanatory variables are included in the modeling process.

REFERENCES

Barden, C. L., and Bramlage, W. J. 1994. Relationships of antioxidants in apple peel to changes in alpha-farnesene and conjugated trienes during storage, and to superficial scald development after storage. *Postharvest Biol. Technol.* 4: 23–33.

Bramlage, W. J., Drake, M., and Lord, W. J. 1980. The influence of mineral nutrition on quality and storage performance of pome fruits grown in North America, Ch. 4, in *Mineral Nutrition of Fruit Trees*, D. Atkinson, J.E. Jackson, R.O. Sharples and W.M. Waller, Eds., Butterworths, Sevenoaks, U.K., pp. 29–39.

Christoffersen, R., Warm. E., and Laties, G. C. 1982. Gene expression during fruit ripening in avocado. *Pl.* 155: 52–57.

De Jager, A., Johnson, D. S., and Hohn, E. (Eds.). 1996. Determination and prediction of optimum harvest date of apples and pears. Proceedings of an EC COST 94 meeting of the Working Group on optimum harvest date. Luxembourg: Office for Official Publications of the European Communities.

Fidler, J. C. 1973. Conditions of storage, Part I, in *The Biology of Apple and Pear Storage*, J. C. Fidler, B. G. Wilkinson, K. L. Edney and R. O. Sharples, Eds., Research Review No. 3, Commonwealth Bureau of Horticulture and Plantation Crops, Commonwealth Agricultural Bureau, Central Sales, Farnham Road, Slough SL2 3BN, U.K., pp. 3–61.

Grierson, D., Slater, A., Speirs, J., and Tucker, G. A. 1985. The appearance of polygalacturonase mRNA in tomatoes: one of a series of changes in gene expression during development and ripening. *Pl.* 163: 263–271.

Hardenburg, R. E., Watada, A. E., and Wang, C. Y. (Eds.). 1986. *The Commercial Storage of Fruits, Vegetables, and Florist and Nursery Stocks.* Agriculture Handbook No. 66 (revised). U.S. Department of Agriculture.

Jackson, J. E., Sharples, R. O., and Palmer, J. W. 1971. The influence of shade and within-tree position on apple fruit size, colour and storage quality. *J. Hort. Sci.* 46: 277–287.

Johnson, D. S. 1989. Mineral composition in relation to storage quality of UK apples. I-setting the standards. *5th Proc. C. A. Research Conference,* Wenatchee, Washington, U.S. I: 31–44.

Johnson, D. S. 1994a. Interactions between apple cultivars and storage conditions with special reference to controlled atmospheres. *Aspects Appl. Biol.* 39: 87–94.

Johnson, D. S. 1994b. Influence of time of flower and fruit thinning on the firmness of "Cox's Orange Pippin" apples at harvest and after storage. *J. Hort. Sci.* 69: 197–203.

Johnson, D. S. 1995. Effect of flower and fruitlet thinning on the maturity of Cox's Orange Pippin apples at harvest. *J. Hort. Sci.* 70: 541–548.

Johnson, D. S. and Ridout, M. S. 1993. Nutritional and meteorological data to predict scald in Bramley's Seedling apples. *Tree Fruit Postharvest Journal.* Washington State University, 4(2): 13–15.

Johnson, D. S. and Ridout, M. S. 1998. Prediction of storage quality of "Cox's Orange Pippin" apples from nutritional and meteorological data using multiple regression models selected by cross validation. *J. Hort. Sci. Biotech.* 73: 622–630.

Johnson, D. S., Bailey, T. C., and Ridout, M. S. 1989. Meteorological and nutritional data to predict low temperature breakdown in stored Bramley's Seedling apples. *Acta Hort.* 258: 455–464.

Knee, M. and Farman, D. 1989. Sources of variation in the quality of CA-stored apples. *5th Proc. C. A. Research Conference,* Wenatchee, Washington, U.S. I: 255–261.

Knee, M., Proctor, F., and Dover, C. J. 1985. The technology of ethylene control: use and removal in post-harvest handling of horticultural commodities. *Ann. Appl. Biol.* 107: 581–595.

Lidster, P. D., Blanpied, G. D. and Prange, R. K. (Eds.). 1990. *Controlled-Atmosphere Disorders of Commercial Fruits and Vegetables.* Agriculture Canada Publication 1847/E. Communications Branch, Agriculture Canada, Ottawa, Ontario.

Perring, M. A. 1986. Incidence of bitter pit in relation to the calcium content of apples: problems and paradoxes, a review. *J. Sci. Fd. Agric.* 37: 591–606.

Richardson, A. 1986. The effects of herbicide soil management systems and nitrogen fertiliser on the eating quality of Cox's Orange Pippin apples. *J. Hort. Sci.* 61(4): 447–456.

Sharples, R. O. 1973. Orchard and climatic factors, Part IV, in *The Biology of Apple and Pear Storage*, J. C. Fidler, B. G. Wilkinson, K. L. Edney and R. O. Sharples, Eds. Research Review No. 3, Commonwealth Bureau of Horticulture and Plantation Crops, Commonwealth Agricultural Bureaux, Central Sales, Farnham Road, Slough SL2 3BN, U.K., pp. 175–225.

Sharples, R. O. 1980. The influence of orchard nutrition on the storage quality of apples and pears grown in the United Kingdom, Ch. 3, in *Mineral Nutrition of Fruit Trees*, D. Atkinson, J.E. Jackson, R.O. Sharples and W.M. Waller, Eds., Butterworths, Sevenoaks, U.K., pp. 17–28.

Sharples, R. O. 1982. Effects of ultra-low oxygen conditions on the storage quality of English Cox's Orange Pippin apples. *3rd Proc. C. A. Research Conference*, Corvallis, U.S., 131–138.

Silsby, K. J. 1993. Identifying apple harvest windows. *6th Proc. C. A. Research Conference*, Ithaca, New York, U.S. 2: 554–563.

Smith. S. M., and Stow, J. R. 1985. Variation in storage and eating quality of clones of Cox's Orange Pippin apples. *J. Hort. Sci.* 60: 297–303.

Stow, J., Alston, F., Hatfield, S., and Genge, P. 1993. New apple selections with inherently low ethene production. *Acta Hort.* 326: 85–92.

Tucker, G. A. 1994. Role of genetic variation in postharvest storage and processing. *Aspects Appl. Biol.* 39: 77–85.

Waller, W. M. 1980. Use of apple analysis, Ch. 66, in *Mineral Nutrition of Fruit Trees*, D. Atkinson, J. E. Jackson, R. O. Sharples and W. M. Waller, Eds., Butterworths, Sevenoaks, U.K., pp. 383–398.

Wilkinson, B. G. and Fidler, J. C. 1973. Physiological disorders, Part II, in *The Biology of Apple and Pear Storage*, J. C. Fidler, B. G. Wilkinson, K. L. Edney and R. O. Sharples, eds. Research Review No. 3, Commonwealth Bureau of Horticulture and Plantation Crops, Commonwealth Agricultural Bureaux, Central Sales, Farnham Road, Slough SL2 3BN, U.K., pp. 65–131.

CHAPTER 5

Environmental Effects on Product Quality

MONIKA SCHREINER
SUSANNE HUYSKENS-KEIL
ANGELIKA KRUMBEIN
ILONA SCHONHOF
MANFRED LINKE

PRODUCT quality as determined by the ultimate consumer is affected by both pre- and postharvest factors. The interaction between single pre- and postharvest variables as they affect overall quality of vegetable products will be described using climatic and storage effects as an example. The objective of the research described in this chapter was the development of a basis for quality management for vegetable crops.

QUALITY DETERMINATION AND QUALITY EVALUATION

Due to surplus production of many horticultural products, competition has led to low prices and a demand for high quality. Therefore, in the past few years quality evaluation and quality control of horticultural products, associated with inspection and monitoring of product quality, has become increasingly important. Strength of competition, when combined with an increased demand for conservation of genetic resources and greater health consciousness, has added to the complexity. This complexity is associated with quality evaluation of fresh fruits and vegetables. Single product attributes do not sufficiently describe the overall quality of a product. Product quality should be assessed by a combination of several characteristic product attributes (Huyskens-Keil, 1996). Consequently, an integral quality value should be developed by summarizing and weighing specific external and internal product attributes.

While it may be easy to collect a comprehensive list of product quality attributes, ascertaining their relative importance is difficult.

Figure 5.1 reveals four groups of product attributes: sensory attributes, bioactive substances, essential nutritive compounds and undesirable attributes. The attributes within each group will be summarized and

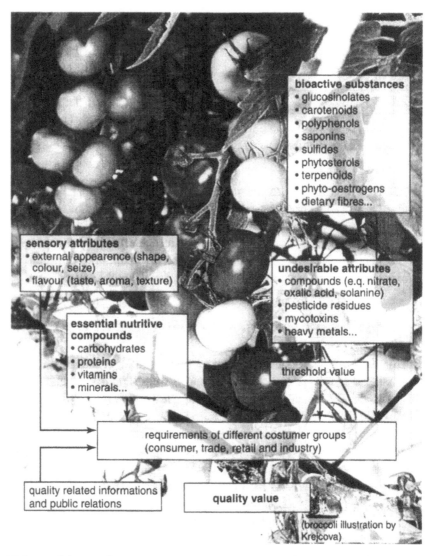

FIGURE 5.1 Overall quality evaluation of vegetable products by summarizing and weighting the several groups of product attributes.

weighted according to the requirements of the different customer groups (consumer, trade, retail and industry). Weighting factors for each group of product attributes can be determined by estimations of an expert panel (van Kooten and Peppelenbos, 1993; Molnar, 1996) or by surveys of each customer group, e.g., consumer surveys (Schonhof et al., 1997; Lennernäs et al., 1997). Further attribute groups might be included, e.g., extrinsic attributes. If the threshold value for undesirable attributes is exceeded, the product will have no market capability (Schreiner et al., 1996).

At present, quality attributes of harvested plant products that are established in official quality standards—like those of the European Community or the USDA Standards—are based on the Codex Alimentarius of the Food and Agriculture Organization of the United Nations (FAO) and the World Health Organization (WHO). In these official standards product quality is primarily based on a subjective assessment of only external product attributes, e.g., whole, firm, clean and fresh in appearence. However, in response to consumer demand there is an increased emphasis on internal product quality, e.g., flavor and health benefits. With increasing health consciousness, additional laws and regulations are being established to reduce undesirable and toxic product attributes (e.g., nitrate). More recently the bioactive substances like glucosinolates, carotenoids, polyphenols, saponins, and different fractions of dietary fibers are generating interest (Figure 5.1) (Schonhof and Krumbein, 1996, 1997; Schreiner et al., 1998). Changes in consumer awareness of health, nutritional, and ecological aspects product quality lead to continual change in the market place. Thus quality cannot be considered fixed but is dependent on consumer perception and time.

EFFECT OF PREHARVEST CLIMATIC CONDITIONS ON BIOACTIVE SUBSTANCES

Due to the health-promoting effect of carotenoids, the influence of air temperature and photosynthetic photon flux density (PPFD) during the production period on the content of β-carotene and lutein has been investigated. The content of β-carotene is dependent on the daily mean temperature, e.g., broccoli cultivar 'Emperor' (Figure 5.2).

Daily mean temperatures below 16.5°C led to an obviously higher β-carotene concentration as daily mean temperatures above this temperature level decrease β-carotene. The PPFD had no effect on the β-carotene or lutein content in broccoli (Figure 5.3).

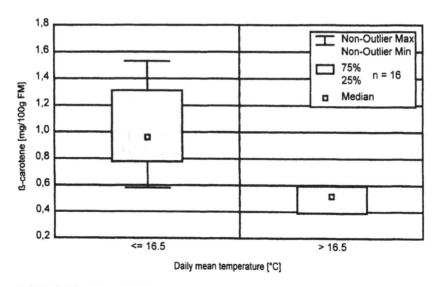

FIGURE 5.2 Effect of daily mean temperature on β-carotene content of broccoli 'Emperor.'

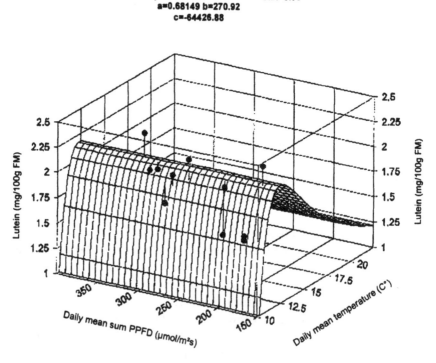

FIGURE 5.3 Correlation between PPFD, temperature and lutein of broccoli 'Emperor.'

Figure 5.3 shows the functional correlation between the lutein content on one hand and PPFD and daily mean temperature on the other hand. The lutein concentration in broccoli reached a maximum at a daily mean temperature of nearly 15°C. These results indicate that β-carotene and lutein biosynthesis is strongly dependent on temperature. Previous investigations support the present results: Carotenoids have been reported to be synthesized in darkness in carrot roots or harvested tomatoes (Gross, 1991).

Another group of bioactive substances are the glucosinolates that mainly occur in cruciferous plants like broccoli (Fenwick and Heany, 1983; Zhang et al., 1992). Glucoraphanin is a main glucosinolate in broccoli, which is a precursor of sulforaphane and described as a protection compound for carcinogenesis (Fahey et al., 1997). Consumers show an increased demand for health benefits in broccoli (Schonhof et al., 1998).

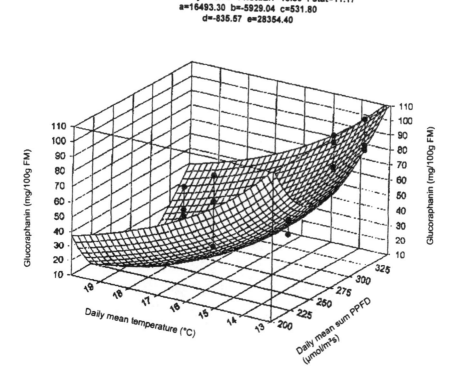

FIGURE 5.4 Correlation between PPFD, temperature and glucoraphanin of broccoli 'Emperor.'

Therefore, the bioactive glucosinolates (glucoraphanin and indolyl glucosinolates) play an important role for broccoli quality. The content of glucoraphanin increased with decreasing daily mean temperature. Additionally, the daily mean of PPFD at nearly 275 μmol m^{-2} s^{-1} led to a minimum of glucoraphanin content at every temperature level (Figure 5.4). Production of glucosinolates and carotenoids is strongly temperature dependent with an increase in temperature inhibiting biosynthesis of carotenoids and glucosinolates.

INTERACTION OF CLIMATIC CONDITIONS
IN PRE- AND POSTHARVEST

According to our results, varying weather conditions and agro management modified the chemical composition of vegetables, that is, cutin accumulation in the cuticula of the leaves and suberin embedded in the secondary tissue of roots and tubers, which directly influences the transpiration rate and indirectly influences the external appearance of the product. The alteration rate of product compounds due to preharvest factors strongly determines postharvest behavior and shelf life.

The seasonal transpiration behavior in postharvest storage will be illustrated by studies on small radish. Figure 5.5 shows the seasonal tran-

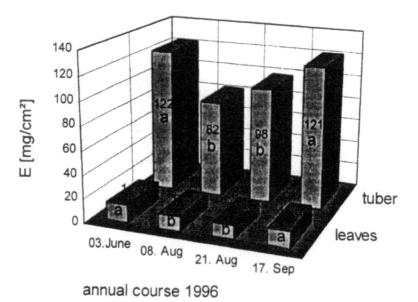

FIGURE 5.5 Seasonal transpiration rate of small radish separated in leaves and tubers.

spiration rate of small radish separated in leaves and tubers. Postharvest storage conditions, weight and dimensions of the single small radishes were nearly constant. Transpiration of the tubers and the leaves showed a pronounced seasonal effect. In comparison to early summer and autumn, the transpiration rate of both product parts in summer was lower and attributed to an increased cutin and suberin synthesis. The rising cutin production was induced by the low relative air humidity (nearly 30%) during the growth period. In a similar way, increasing soil temperatures associated with lower soil humidity might lead to a stronger suberin insertion in the external tissue layer of the tubers. In comparison with the tubers, the leaves showed a reduced transpiration rate, which might be explained by a better adaptation to high water potential differences.

EFFECT OF POSTHARVEST STORAGE CONDITIONS ON TEXTURAL PROPERTIES

The altering gas exchange (i.e., changes in respiration, transpiration and ethylene production) was found to be closely correlated with changes in the textural properties of the cell wall during development and ripening (Kader, 1987; Dick and Labavitch, 1989; Huyskens, 1991; Mendlinger et al., 1992; Sonego et al., 1995). Softening is dependent on the physiological stage of the fruit or vegetable and results from a transformation of insoluble pectic substances to water-soluble pectin leading to losses of quality (Huyskens-Keil et al., 1998).

The strong relationship between the change of total pectic substances and respiration rate during postharvest storage and changes in water content, related to the fresh matter at harvest time, are presented in Figure 5.6 for lettuce. The data were estimated by a regression analysis, where respiration rate was used as an independent variable among other variables. Respiration and the decomposition of total pectic substances depended on the water status of the product. A decrease of pectic substances occurred at a low water content of lettuce (65%) (Schreiner et al., 1995). Reduced decomposition of pectic substances with rising respiration may be due to the enhanced energy demand for detaching monosaccharides linked to pectic substances and for utilizing them as additional respiration substrates. At a high water status of lettuce (87%, 85%), there was no decomposition of pectic substances. The negative values of pectin degradation at higher water status indicate preferred respiration of storage carbohydrates.

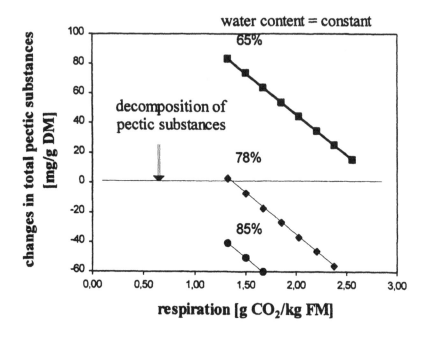

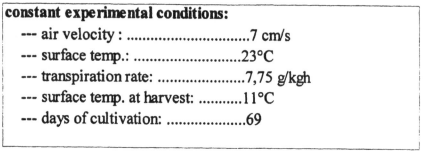

FIGURE 5.6 Relative changes in total pectic substances during postharvest treatment in lettuce.

EFFECT OF POSTHARVEST STORAGE CONDITIONS ON SHELF LIFE

Improving quality is associated with quality control. Quality control of horticultural products in postharvest can be achieved via determination of product-specific thresholds. These thresholds define, for example, the permitted range of storage-mediated changes in the product resulting in a nonvisual external quality loss. Leafy vegetables or vegetables purchased with leaves are characterized by a locally differentiated wilting behavior in postharvest. Bunched carrots and small radishes show their first wilting symptoms on the leaves. For assessing product-specific thresholds of bunched carrots and small radishes, the critical water loss was determined by the shift from turgidity to the appearance of slight wilting symptoms on the leaves. The maximum permissible water loss is 4% for bunched carrots and 5% for small radishes (Kays, 1991). By knowing the maximum acceptable water loss of bunched carrots and small radishes and knowing the postharvest storage conditions, it is possible to predict the time unit for shelf life. The product-specific threshold represents the end of market life. This process is shown in the Figure 5.7 using bunched carrots as an illustration. The relative air humidity was held to be constant at 80%, while the air temperature varied from 0°C to 25°C. In this case the specific product threshold was defined by the critical water content (86%), because the entire relative water concentration of harvested carrots with leaves was 90%. It is obvious from this example that in the case of 25°C air temperature, the threshold was reached in four hours. In contrast air temperature of 5°C caused an shelf-life extension of about 16 hours.

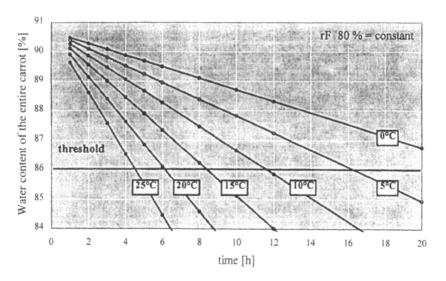

FIGURE 5.7 Shelf-life prediction for carrots.

CONCLUSIONS

The results presented here show changes of quality attributes with respect to preharvest climatic factors and postharvest storage conditions. Using this integrated approach across pre- and postharvest periods, a comprehensive quality evaluation is being developed. Further investigations will be conducted to confirm the correlations found between climatic and storage effects as they relate to overall quality changes in different horticultural products.

REFERENCES

Dick, A. and Labavitch, J. 1989. Cell wall metabolism in ripening fruits. *Plant Physiol.* 89: 1394–1400.

Fahey, J., Zhang, Y., and Talalay, P. 1997. Broccoli sprouts: An exceptionally rich source of inducers of enzymes that protect against carcinogens. *Proc. Natl. Acad. Sci. USA* 94: 10367–10372.

Fenwick, G. and Heany, R. 1983. Glucosinolates and their breakdown products in cruciferous crops, foods and feedingstuffs. *Food Chemistry* 11: 249–271.

Gross, J. 1991. *Pigments in Vegetables—Chlorophylls and Carotenoids*, Van Nostrand Reinhold, New York.

Huyskens, S. 1991. Morphological, physiological and biochemical aspects in the cultivation of two pantropical cucurbits *Luffa acutangula* (L.) Roxb. and *Momordica charantia* L. Dissertation, Universität Bonn.

Huyskens-Keil, S. 1996. Qualität von Obst und Gemüse—Möglichkeiten und Grenzen einer integralen Bewertung. *Bornimer Agrartechnische Berichte* 8: 9–15.

Huyskens-Keil, S., Schreiner, M., and Widell, St. 1998. Qualitätssicherung bei Radies. *Taspo-Gartenbaumagazin* 2: 40–41.

Kader, A. 1987. Respiration and gas exchange of vegetables, in *Postharvest Physiology of Vegetables*, J. Weichmann, Ed., Marcel Decker, Inc., New York and Basel, pp. 25–45.

Kays, S. 1991. *Postharvest Physiology and Handling of Perishable Plant Products*, Van Nostrand-Reinhold, New York.

Lennernäs, M., Fjellström, C., Giachetti, I., Schmitt, A., Remaut de Winter, A., and Kearny, M. 1997. Influences on food choice perceived to be important by nationally representative samples of adults in the European union. *Eur. J. Clin. Nutr.* 51, Suppl. 2: 8–15.

Mendlinger, S., Benzioni, A., Huyskens, S., and Ventura, M. 1992. Fruit development and postharvest physiology of *Cucumis metuliferus* Mey., a new crop plant. *J. Hort. Sci* 67(4): 489–493.

Molnar, P. 1996. A model for overall description of food quality. *Food Qual. Pref.* 6: 185–190.

Schonhof, I. and Krumbein, A. 1996. Carotinoide in Brokkoli und ihre Variabilität in Abhängigkeit von Sorte und Anbauzeitraum. 16. *Tagung der Deutschen Gesellschaft für Qualitätsforschung, Kiel:* 37–40.

Schonhof, I. and Krumbein, A. 1997. Einfluß des Anbauverfahrens Kulturschutzabdeckung auf die Qualität von Brokkoli. 17. *Tagung der Deutschen Gesellschaft für Qualitätsforschung, Wädenswil:* 279–282.

Schonhof, I., Brückner, B., and Röger, B. 1997. Innere Werte von Möhren rangieren beim Verbraucher vor den äußeren. *Taspo-Gartenbaumagazin* 2: 41–42.

Schonhof, I., Brückner, B., Krumbein, A., and Gutezeit, B. 1998. Was Verbraucher an Brokkoli schätzen. *Deutscher Gartenbau* (32): 43–45.

Schreiner, M., Linke, M., and Huyskens, S. 1995. Measurements of produce responses on climatic impacts in postharvest. Reprint of the 1st IFAC/CIGR/EURAGENG/ISHS workshop on control applications in postharvest and processing technology, 1–2 June 1995, Oostende (Belgium): 193–200.

Schreiner, M., Linke, M., and Huyskens, S. 1996. Nacherntequalitätskriterien bei Kopfsalat (*Lactuca sativa* L. var. *capitata* L.). *Bornimer Agrartechnische Berichte* 8: 33–50.

Schreiner, M., Schonhof, I., and Krumbein, A. 1998. Eine neue Dimension der Produktqualität—Bioaktive Substanzen im Gemüse. *Gemüse* 2: 80–84.

Sonego, L., Ben-Arie, R., Raynal, J., and Pech, J. 1995. Biochemical and physical evaluation of textural characteristics of nectarines exhibiting woolly breakdown: NMR imaging, X-ray computed tomography and pectin composition. *Postharvest Biol. Technol.* 5(3): 187–198.

Van Kooten, O. and Peppelenbos, H. 1993. Predicting the potential to form roots on chrysanthemum cuttings during storage in controlled atmospheres. *Proceedings of the 6th international controlled atmosphere reasearch conference*, Cornell University, Ithaca, New York, pp. 610–619.

Zhang, Y., Talalay, P., Cho, C., and Posner, G. 1992. A major inducer of anticarcinogenic protective enzymes from broccoli: isolation and elucidation of structure. *Proc. Natl. Sci. USA* 89: 2399–2403.

CHAPTER 6

Postharvest Handling and Storage of Vegetables

TORSTEN NILSSON

INTRODUCTION

POSTHARVEST handling and storage of vegetables is a necessary measure in order to supply consumers, caterers and processors with fresh produce independent of the distance in time and space between grower and consumer. The vegetable market comprises a broad assortment from fresh produce and minimally processed refrigerated vegetables (Wiley, 1994) to canned, frozen, fermented or just pre-cooked products treated in order to maintain shelf life and quality over a longer period of time (Arthey and Dennis, 1991). This chapter will focus on storage and handling of fresh and minimally processed, refrigerated vegetables.

Vegetables comprise a wide range of plant organs from those undergoing rapid preharvest growth (asparagus, broccoli, lettuce, cauliflower), which makes them perishable with a limited shelf life, fleshy fruits (tomato, cucumber, sweet pepper) to storage organs (onion, carrot, celeriac, beetroot) designed for survival to the next growing season. Those belonging to the first two groups are not intended to continue their ontogenesis after harvest and therefore senesce rapidly due to inability to maintain homeostasis. In the third group, mainly consisting of plants with a biennial life cycle, growth ceases in the autumn with the storage organ entering an imposed or true dormant period with diminished metabolism. Normally, storage organs are carrying meristematic tissue, which undergoes vernalization during storage. Regrowth with a con-

96

comitant senescence of the storage tissue therefore often limits the duration of storage.

These differences in pre- and postharvest physiology among vegetables originally restricted the assortment of vegetables available to many consumers apart from those living in regions with a climate allowing vegetable growing all year round. Improved techniques for distribution of fresh vegetables over long distances have contributed to a broad assortment available independent of season. The need for long-term storage (several months) in order to secure supply of certain vegetables from one season to the next is thus merely restricted to the temperate zone. Evolving market demands, especially in many industrial countries, for accessibility of all kinds of fresh vegetables independent of season and geographic localization have made proper postharvest handling crucial for reduction of losses between grower and consumer. To achieve this goal the communication barriers along the postharvest chain must be broken down.

Independent of the time span from harvest to consumption, all vegetables are perishable. Every effort aimed at retarding quality deterioration due to senescence or desiccation or at encouraging and controlling fruit ripening during the postharvest chain must rely on a continuously improved knowledge about the physiology of vegetables as living produce and the interaction between environment and produce as it influences quality and shelf life. This is an ongoing challenge to growers, wholesalers, distributors and retailers. However, maintaining product quality during storage and subsequent shelf life in a way that will satisfy consumers' demands implies today an integration of both crop production amendments, stage of development at harvest as well as a proper, safe and healthy postharvest handling of the produce. This chapter is an attempt to elucidate the present knowledge about how the living plant tissues of vegetables interact with their environment during storage and handling and implications on the quality of the produce when reaching the ultimate judge—the consumer.

PHYSIOLOGICAL BASIS FOR POSTHARVEST QUALITY MAINTENANCE

Harvesting plant organs for human consumption involves several stresses to the plant tissue. Photosynthesis normally ceases at harvest, resulting in a changed allocation of carbon with new sources and sinks. Water uptake through the roots is terminated, but moisture loss due to transpiration continues. Translocation of plant hormones from the roots (e.g., cytokinins) is also interrupted but bruises and lesions may introduce stress metabolism in the wounded tissue initiating synthesis of ethylene with an enhanced respiration. Wounding also facilitates entrance of various pathogenes resulting in postharvest decay. Taken together, vegetables must be harvested with special care in order to avoid deleterious postharvest processes resulting in severe reduction of the keeping quality and losses of such freshness attributes as appearance, texture and flavor.

Vegetables make up a very heterogeneous group of food from plants compared with cereals. The type of plant organ harvested and its role in the ontogeny of the growing plant has a strong influence on storage potential and shelf life. The physiological basis for maintaining quality during postharvest handling is therefore closely related to the ability of the plant organ harvested to maintain metabolic homeostasis for a period of suitable length provided the storage conditions are close to optimal. Consequently, vegetables have been divided into subgroups with more or less similar behavior and storage potential.

Leaves, Stems, Flower Buds and Inflorescences

Most vegetables within this group are harvested at a stage of horticultural maturity (Watada et al., 1984) determined by consumer demands or suitability for processing rather than the degree of physiological maturity and suitability for postharvest storage and subsequent shelf life. Vegetables harvested can consist of the whole overground plant (lettuce, Chinese cabbage, leek), a section of the plant (broccoli, cauliflower) or single leaves (parsley, spinach). Due to the large proportion of young tissue, preharvest respiration and overall cell metabolism are on a high level and continue in the harvested plant material. The stage of development at harvest is often crucial due to ongoing quality-reducing processes, e.g., cell wall lignification in asparagus spears or accumulation of starch in green peas and sweet corn. The epidermis or periderm exerts little protection against postharvest water loss. Rapid precooling of the produce immediately after harvest, aimed at slowing down cell metabolism and transpiration, followed by an unbroken cool chain is a prerequisite for achieving an acceptable postharvest quality retention. However, since the tissue of these vegetables is programmed for postharvest senescence rather than dormancy, the inherent storage potential is limited due to progressive senescence. Shelf life of such highly perishable vegetables is therefore limited from a few days to a few weeks.

Fruits and Pulses

Vegetables as a group of food plants also include some fleshy fruits (tomato, cucumber) and immature fruits (e.g., peas and bean pulses). Compared with most vegetables, postharvest storage of fleshy fruits shows both similarities and differences. The function of a fruit—quite different from that of vegetables—is to protect the developing seeds against predators and later to attract the same predators (animals or humans) in order to secure seed distribution. This means that the physiological basis for storage and shelf life of fleshy fruits is closely associated with their maturation and subsequent ripening and senescence.

Fruit can be classified either as climacteric or nonclimacteric due to the pattern of respiration and ethylene production during ripening. Climacteric fruit are characterized by a transient increased rate of respiration and ethylene production with the magnitude varying considerably between fruit species. Fruit with a narrow respiratory peak may change from unripe to overripe within a few days when kept at room temperature, while fruit with a wider peak may remain ripe for several days or weeks. During the growing period, most climacteric fruit accumulate starch, which is then degraded to soluble sugars during ripening providing both sweetness to the fruit and substrate for respiration and synthesis of, e.g., pigments and aroma compounds.

Nonclimacteric fruit ripen without any increase in the rate of respiration and ethylene production. Since these fruits do not accumulate starch, their sweetness is dependent on the ability of the plant or tree to export carbohydrates to the fruit during the period preceding harvest. This means that harvesting of nonclimacteric fruits must be delayed until the fruit has reached full ripeness. While most climacteric fruits intended for storage or distribution over long distances must be harvested preclimacteric, those with a nonclimacteric ripening may be harvested ripe and often stored for several weeks without any change in eating quality.

Biennial Vegetables

Vegetables belonging to this group normally complete their life cycle over a two-year period but low temperatures during the seedling stage may, depending on species and cultivar, result in vernalization and sprouting already in the first year. Normally, some kind of storage organ is developed in the first season of growth followed by flowering and seed formation with subsequent tissue death in the second year. Morphologically, biennial vegetables can differ considerably—from a storage root (carrot), a combined hypocotyl and root (beetroot), a rhizome (Jerusalem artichoke) to a large number of folded leaves emanating from the same central core carrying the apical meristem (winter white cabbage) or swollen leaf bases emanating from a dwarf stem (onion bulbs). With ceasing growth in the autumn, the storage organ enters into an imposed or true dormant period with diminished metabolic rate, which makes them adapted for long-term storage.

INFLUENCES OF THE ENVIRONMENT

Temperature

Temperature is one of the most important factors determining the keeping quality of horticultural produce due to its influence on physical, biochemical and inductive processes. Lowering the temperature reduces respiration as well as other metabolic processes, and therefore delays senescence. There is a linear relationship between the logarithm of oxygen consumption rate and tissue temperature. The effect of a 10°C increased temperature on respiration, the Q_{10} value, averages 2.0 to 2.5 within the 5 to 25°C temperature range. At temperatures > 25°C, Q_{10} generally decreases probably due to a gradual denaturation of enzymes and limited diffusion rate of oxygen from surrounding air to the mitochondria.

With decreasing temperature, the kinetic energy of water molecules diminishes, resulting in a reduced vaporization rate of liquid water with a lower transpiration and water loss. Maintaining freshness and quality during the postharvest chain is thus mainly a question of temperature control but the optimum temperatures for keeping quality of various products vary considerably. A number of plants of tropical or subtropical origin develop chilling injury when stored at temperatures below a certain limit (Saltveit and Morris, 1990). Among vegetables affected by chilling injury, the tomato group consisting of squash, cucumber, water melon, aubergine, and tomato has a lower limit at 12°C while the potato group consisting of snap beans, sweet pepper, and potato has a lower limit at 5–8°C. Vegetables originating from the temperate zone are preferentially stored at a temperature between zero and 2°C. All produce affected by chilling injury must therefore be kept at temperatures that limit shelf life due to the high metabolic rate compared with chilling tolerant produce.

Rapid precooling close to harvest and an unbroken cool chain contribute to a prolonged shelf life but the benefit of precooling is dependent on tissue temperature at harvest, respiratory activity of the vegetable and the possibilities for maintaining the optimum storage temperature along the entire postharvest chain. For production of minimally processed, refrigerated (MPR) vegetables, it is essential to start with precooled material and maintain a low temperature during chopping, washing, packaging and distribution. For such ready-to-eat vegetables maintenance of a stable temperature is also crucial for avoiding microbial spoilage, and therefore the key to success for consumer-packed MPR products (Ahvenainen, 1996).

Methods for prediction of shelf life/storage life at constant temperatures have been published for many products (Thorne and Meffert, 1979). Most shelf-life studies (Labuza, 1982; Shewfelt, 1986) of fresh vegetables assume that the rate of deterioration according to time is independent of the quality level, which gives zero order reactions and linear relationships between number of days from harvest and quality deterioration (van Beek et al., 1985). However, during distribution and retailing most produce is subjected to irregular temperature regimes. Is it then possible to estimate the amount of remaining quality at the time a product is delivered to the retailer?

Applications of the Time-Temperature-Tolerance method (van Arsdel, 1957) rest upon the assumption that changes in harvested products are additive and commutative and that there are no effects of the temperature change itself. "Additivity" means that the total loss of quality is the sum of the amount of quality lost at each temperature whereas "commutativity" means that the total amount of quality lost is independent of the order of the various temperatures exposed to the product. In an investigation with tomatoes cv. 'Nomato,' Thorne and Alvarez (1982) were able to demonstrate that color and firmness changes were both additive and commutative and could be used to predict the storage life of tomato fruits under any fluctuating temperature conditions between 12 and 27°C.

Low temperatures enhance the storability and shelf life of most vegetables as long as the temperature is above the threshold level for chilling injury or freezing damages of the tissue. For biennial vegetables long-term storage at a temperature just above zero will lead to vernalization with subsequent regrowth and sprouting when the produce is exposed to long-day conditions during retailing. A similar situation is often present in Chinese cabbage where vernalization may occur in the field with a rapid development of an inflorescence inside the cabbage head (Elers and Wiebe, 1984a, b). Nondestructive methods for detection of such hidden defects established preharvest or during storage with appearance in the kitchen need to be improved.

For most vegetables, optimum temperatures for growth and development are far above those most suitable for storage and an extended shelf life. Are such low temperatures beneficial to quality retention and improvement besides freshness? Lowering the tissue temperature will influence the activation energy of enzyme-catalyzed reactions, with certain reactions being more affected than others leading to imbalances in cell metabolism. Low-temperature sweetening in potatoes is one well-known example caused by the temperature sensitivity of some of the glycolytic enzymes (ap Rees et al., 1988). Transfer of potatoes to a higher temperature before processing may reestablish normal cell metabolism and avoid undesired browning due to Maillard reactions in fried potato products. Since freshness and turgor are essential for quality of vegetables, few if any investigations have been made until present about the conceivable effects of low-temperature storage on the taste, texture and flavor of the produce.

An increasing concern among consumers about chemical treatments applied pre- or postharvest to fruit and vegetables to control insects, diseases, physiological disorders and regrowth has called upon less harmful alternatives. Postharvest heat treatments aimed at insect disinfestation and disease control can also alter the senescence of fruit by influencing the rate of protein synthesis, softening, chlorophyl loss, respiration and ethylene synthesis (Paull, 1990; Klein and Lurie, 1991). Heat treatments at 38–46°C for 12 hr to 4 days or short-term heat treatment for up to 60 min at 45–55°C by exposure to hot air, vapor heat or immersion in hot water (Paull, 1990) have been used for slowing the ripening of climacteric fruits resulting in a longer shelf life. Hot water treatments of broccoli florets at a range of temperatures (43–55°C) with a duration from 10 sec to 30 min have been tested in order to delay yellowing and extend shelf life (Forney, 1995; Tian et al., 1996, 1997). Heat treatment inhibited wound-induced ethylene production with delayed degreening compared with the control. Ethylene production plays an important role in broccoli yellowing probably due to the climacteric behavior of this inflorescence tissue (Tian et al., 1994). The effect of heat treatment on yellowing could be mediated via a transient inhibition of ethylene synthesis similar to results reported for heat treated tomatoes (Biggs et al., 1988). The potential benefit of heat treatment to delay senescence in non-fruit and non-flower vegetables should be a fertile field for future research.

Humidity

Moisture loss from harvested vegetables causes changes in structure, texture and appearance (Woods, 1990). Reduction of water loss during the postharvest period is therefore critical for maintaining freshness and quality. During growth, water taken up by the roots serves as a transport medium for minerals in the xylem and organic compounds (carbohydrates, amino acids) in the phloem. Water surplus escapes as water vapor from the entire plant surface but mainly through the stomata on the leaves, and regulation of the rate of transpiration by stomata is well known. After harvest, water supply from the roots is terminated and the ability to withstand water loss is now mainly dependent on the water vapor pressure deficit between produce and surrounding air as well as the surface transfer resistance to water vapor movement since stomata are closed. The former could be considered as a pure physical phenomenon while the latter is determined by shape as well as the resistance to water movement inside the tissue and the surface transfer resistance normally defined as the transpiration coefficient (Sastry et al., 1978).

Plants grown under arid conditions build up a strong inherent resistance to water loss compared with plants grown under more humid conditions. Preharvest conditions thus exert a strong influence on the ability to retain turgor and freshness during the postharvest period. Two routes are available for moisture through the surface, evaporation from the produce surface and transpiration via small openings between the epidermis cells. The plant surface could consist of a thin layer of lightly suberized living cells, a suberized wall of dead cells or a wall of living cells covered by a waxy cuticle. Surface composition is not constant for a particular vegetable but may change both during growth and development and due to postharvest senescence. Being the barrier protecting the tissue against environmental forces, the evaporating surface exerts a strong influence on the resistance to moisture loss from a produce. Apart from leafy vegetables, only a very small proportion of the plant surface may consist of apertures connecting intercellular space with the surrounding air for transfer of oxygen and carbon dioxide. Water vapor movement is reported to be low through these apertures (Burton, 1982). This may also explain why micro-perforated packaging materials protect against water loss without establishing a modified atmosphere inside the cover.

The transpiration coefficient as a measure of surface transfer resistance not only vary considerably between different produce (Sastry et al., 1978) but also for the same produce dependent on internal turgor and external air humidity as well as stage of development. Wrapping of desiccation-sensitive produce with either a plastic film, waxing or application of different edible coatings (Nussinovitch and Lurie, 1995) is therefore a common practice. The multilayered structure of onions and iceberg lettuce is carrying an inherent protection since water transport between leaves occurs via the dwarf stem. The main water loss is therefore from the outer leaves, and as they dry they will form a barrier against further water loss. This desiccation is acceptable for dry onions but for iceberg lettuce wilted outer leaves are not acceptable.

Without any air movement, a boundary layer of water vapor-saturated air is established close to the vegetable that increases the resistance to transpiration. Air movement should therefore be as low as possible but the only possibility for removing respiratory heat is via convection, which implies air movement. Rapid precooling with water vapor-saturated air reduces water loss. Film-wrapped vegetables usually require longer cooling time than unwrapped produce, which could result in water condensation within the package (Kadar et al., 1989). After cooling to the desired temperature, the respiratory heat of the vegetables will maintain a water vapor pressure deficit with continued transpiration even though the air inside the package is saturated.

For those involved in storage, distribution and retailing of vegetables both the influence and measurement of temperature are easy to understand. However, humidity and its measurement is much more difficult to grasp. Air humidity is commonly specified as relative humidity (RH), which is the ratio of water vapor pressure in air to saturation vapor pressure at the same temperature, expressed as a percentage. The use of relative humidity measurements has the disadvantages that as long as produce temperature differs from that of the surrounding air, a water vapor pressure deficit is present between produce and surrounding air even if air humidity approaches saturation.

Although most vegetables have a high water content, differences in the osmotic pressure of the cell sap are present between various species and tissues. The water vapor pressure exerted by the produce in the intercellular space is therefore lower than that of distilled water. If placed in a sealed enclosure under constant temperature, the vegetable will gain or lose water until the humidity inside the enclosure reaches a value that is characteristic of that vegetable at that temperature. This value is called the equilibrium relative humidity (ERH) (Rooke and van den Berg, 1985). Most vegetables have an ERH between 97–98% compared with

75–80% for dry onions. Stored in air with a lower RH, most vegetables lose water with the rate of transpiration dependent on water vapor pressure deficit and the resistance of the tissue to water loss. Storage in air with an RH at 97–98% requires a proper temperature control. Even small temperature changes will result in water condensation that may promote microbial decay. Recommended humidities for individual products are usually a compromise between these two conflicting requirements.

Air Composition

Metabolic processes in plant cells are mainly either an oxidation or a reduction. The availability of oxygen at the reaction center will therefore determine the rate of many catabolic processes that influence quality. Respiration in plants is an oxidative degradation of sugars, organic acids and lipids to produce carbon dioxide and water with release of energy. Modifying the atmosphere around the produce, i.e., lowering the amount of O_2 with an increased amount of CO_2, may lower the metabolism with decreasing CO_2 production and O_2 consumption. The effects of low O_2 and high CO_2 are additive but the optimal concentrations of the two gases in the storage atmosphere vary between vegetables and even between cultivars of the same species (Saltveit, 1989).

Due to the high affinity for O_2 of the terminal oxidazes in the electron transport chain located in the mitochondria, the amount of O_2 in the surrounding air must be reduced below 10%. On the other hand a change to anaerobic respiration will take place if the O_2 concentration approaches 2%. The effects of an increased CO_2 concentration are more difficult to explain since elevated CO_2 normally stimulates respiration in stored roots and bulbs (Weichmann, 1977). High CO_2 levels are reported to inhibit enzymes in the Krebs cycle (Nanos et al., 1994) and decreasing pH of the cell sap.

The benefit of modified-atmosphere storage relies on the application of an optimal storage temperature. Under these circumstances, modified atmospheres may delay quality deterioration and prolong the storage period. Long-term (>6 months) controlled atmosphere (CA) storage is mainly used for storage of climacteric fruit, onions, cabbage and Chinese cabbage. The benefit of CA storage of these vegetables is a delayed vernalization and sprouting, retention of color due to retarded breakdown of chlorophyll and reduced microbial decay (Isenberg et al., 1987). Modified atmosphere (MA) has become a valuable tool for consumer packaging of vegetables (MAP) and mainly minimal processed refrigerated vegetables (MPR). Besides reducing respiration, MA can reduce microbial spoilage, the incidence of physiological disorders, degradation of chlorophyll and enzymatic browning.

Preparation of minimally processed, refrigerated vegetables entails wounding of the plant material. Slicing, chopping, dicing or shredding result in an increased respiration and ethylene production together with exposure of leakage cell content. MAP may reduce enzymatic peroxidation of unsaturated fatty acids catalyzed by lipoxygenase with the formation of aldehydes and ketones resulting in off-flavor. Ethylene production is dependent on oxygen and carbon dioxide is reported to inhibit ethylene action by blockage of the receptors. MAP will therefore delay senescence and chlorophyll degradation induced by ethylene. Discoloration due to formation of brown polymers from oxidation of *o*-diphenols to *o*-quinones catalyzed by the enzyme polyphenol oxidase may also be inhibited by MAP. This enzyme has a low affinity for oxygen but markedly dependent on pH, which is why added citric acid or ascorbic acid to MPR will inhibit enzymatic browning more than MAP alone.

PHYSIOLOGY OF STORAGE

Senescence and Ripening

In plants showing a monocarpic behavior, the ultimate stage of development implies a massive mobilization of nitrogen, carbon and minerals from leaves, roots and stems to the developing seeds, resulting in whole-plant senescence with subsequent death of the remaining tissue. The changes taking place form a genetically programmed sequence with close coordination at the cell and tissue level (Buchanan-Wollaston, 1997). During growth, similar endogenously controlled deteriorative changes may also take place in the older leaves due to unfavorable conditions such as nitrogen deficiency, light limitation or drought. In the senescing leaf, degradation of macromolecules is followed by remobilization of the components to developing parts of the plant. In vegetables composed of both old leaves and young developing leaves, e.g., cabbage, onions and lettuce, postharvest senescence of the outermost leaves will result in a translocation of carbohydrates and amino acids to the younger leaves and the stem but for many harvested vegetables a similar remobilization is not possible. The senescence process seems therefore not to be dependent on concurrent deprivation of the senescing tissue.

If senescence in monocarpic plants represents a genetically programmed developmental phase, does senescence in harvested vegetables show any similarities? Today there is growing agreement that senescence of harvested plant organs proceeds in a similar, genetically controlled manner as in intact plants. This means that although environmental factors may trigger the onset and progress of postharvest senescence, the environment does not regulate it. One difference between natural and induced senescence is that induced senescence may be reversible if stress conditions are removed before senescence has reached a certain point (Smart, 1994).

Senescence in harvested plant organs as vegetables must therefore be considered as a premature, induced process—if left to continue growth on the plant, senescence will not occur after a similar period of time. Both natural and induced senescence proceed in a sequential manner probably with increasing dominance of catabolic processes leading to cellular breakdown. However, postharvest senescence is not a passive decay but rather an actively regulated process. Cells remain viable and show tight metabolic regulation. Mitochondria retain their function since respiration must continue until the end of senescence. The expression of many genes is downregulated in senescing tissue, probably necessary to prevent regeneration of cell components broken down during senescence (Noodén et al., 1997). Concomitantly, there is a coordinated expression of specific senescence genes, and the transcription of such genes seems to be similar in both natural and induced senescence (Buchanan-Wollaston, 1997). The senescence syndrome is composed of several parallel processes such as loss of chlorophyll and dismantling of chloroplasts, degradation of soluble protein to amino acids and generation of oxygen radicals resulting in lipid peroxidation and a general loss of cellular compartmentation leading to tissue breakdown.

The most evident symptom of senescence in harvested vegetables is the loss of green color due to degradation of chlorophyll. Normally, light delays loss of chlorophyll in detached leaves (Okada et al., 1992) but most vegetables are stored in darkness. Cold storage of cabbage over 150 days resulted in a gradual loss of chlorophyll and degradation in products with a color change from green to white (Heaton et al., 1996), which is in agreement with Prange and Lidster (1991), who obtained better chlorophyll retention in cabbage stored at low light intensities compared with darkness. However, few, if any, investigations have described the influence of light on color retention during handling and retailing of vegetables.

Degreening is the most clear indication of ageing and decreasing freshness of vegetables besides loss of turgor and decay. Improved knowledge about the processes that initiate and regulate chlorophyll degradation is therefore of paramount importance in postharvest handling of vegetables to design techniques to prevent or delay the onset of senescence. This is closely connected to the question about duration of freshness after harvest. Chlorophyll degradation is not visible to the eye until senescence is advanced, making color judgements less suitable for evaluation of freshness. Search for a suitable nondestructive "early senescence marker" is therefore of high priority, especially for minimally processed, refrigerated vegetables.

The degradation of chlorophyll is assumed to start with the removal of the phytol chain by the enzyme chlorophyllase resulting in the formation of chlorophyllide. The following steps include replacement of the central magnesium by hydrogen to form pheophorbide, and finally cleavage of the porphyrin ring system by a dioxygenase with formation of colorless low molecular weight compounds stored in the vacuole (Heaton and Marangoni, 1996). Ethylene seems to play an important role both for initiating and coordinating chlorophyll breakdown in some plants, preferably dicots. It is also a common experience that exogenous ethylene enhances degreening of most vegetables (e.g., mixed loads containing both ethylene-producing fruit and vegetables) but the pathway is still unclear. Ethylene is reported to increase chlorophyll degradation with enhanced chlorophyllase activity and accumulation of chlorophyllide in the flavedo of mature green oranges (Amir-Shapira et al., 1987; Trebitsh et al., 1993). Investigations made into breakdown of chlorophyll in harvested vegetables stored in air, air + 10 ppm ethylene or modified atmosphere (Yamauchi and Watada, 1991, 1993, 1998) indicate the presence of different pathways for chlorophyll degradation.

The role of ethylene as an activator of senescence-related genes is unclear. Ethylene is probably involved in plant responses to external signals such as wounding, pathogen and environmental stress but ethylene seems not to activate senescence genes directly (Buchanan-Wollaston, 1997). Senescence of detached leaves was enhanced by wounding with the first peak of ethylene as a response to detachment (Philosoph-Hadas et al., 1989). Ethylene is usually produced at very low levels in vegetative tissue. Wounding induces enzymes in the ethylene biosynthetic pathway, resulting in a burst of ethylene that may occur immediately or some hours after harvest (Abeles et al., 1992). Harvest always introduces wounding stress and it is therefore conceivable that postharvest degradation of chlorophyll is induced by an enhanced stress-mediated ethylene production. Trace amounts of ethylene (less than 0.1 ppm) in the storage atmosphere are physiologically active and therefore promote endogenous ethylene production and senescence. Senescing plant organs are able to produce sufficient amounts of ethylene to accelerate senescence of the entire content in a closed box of vegetables as well as during exposure in a retail cabinet provided the temperature is approaching room temperature. Careful handling in order to reduce mechanical damage (Abeles et al., 1992), misting to maintain turgor (Barth et al., 1992), modified atmosphere packaging of perishable products to inhibit ethylene synthesis and protection from exogenic ethylene sources (Yamauchi and Watada, 1993), and removal of waste or infested products during the whole postharvest chain will therefore reduce degreening and delay senescence.

The involvement of free radicals in the process of leaf senescence is well established (Leshem, 1988). The chloroplast is a main cellular component where reactive oxygen species are generated and inability of the chloroplast to eliminate free radicals may result in an accumulation of lipid peroxidation products. The main cellular components susceptible to damage by free radicals are polyunsaturated fatty acid in membranes, proteins, carbohydrates, nucleic acids and pigments such as chlorophyll and carotenoids. Living systems are protected from active oxygen species by enzymes such as superoxide dismutase, peroxidase and catalase. Besides these enzymatic antioxidants, glutathione, tocopherols, ascorbic acid, carotenoids and other reducing compounds also contribute to the elimination of reactive oxygen species. It has been proposed that initiation of senescence results in generation of oxygen radicals (Leshem, 1988; Dhindsa et al., 1981), with the progress of lipid peroxidation inhibited by the ability of the antioxidant system to quench various oxygen radicals. Recently, Toivonen and Sweeney (1998) have reported that the difference in retention of green color between two broccoli cultivars could be related to the rate of superoxide dismutase and peroxidase activities, which were approximately 30% higher in the cultivar retaining a stable chlorophyll content over four days at 13°C. The chlorophyll breakdown in three edible herbs was related to the oxidative defense system and not to proteolysis, indicating a close relationship between chlorophyll breakdown and lipid oxidation (Philosoph-Hadas et al., 1994).

Mechanisms to prevent peroxidation of membrane lipids are linked to the activity of antioxidant enzymes and endogenous antioxidants (Leshem, 1988). According to this hypothesis, senescence is a consequence of the increasing inability of the tissue to maintain a total antioxidant level necessary for a simultaneous elimination of oxygen radicals. This means that freshness should have a bearing not only on an attractive appearance of the vegetable but also for the food value due to a higher level of antioxidants in the tissue. During recent years several reports have reviewed the epidemiological evidence that increased consumption of fruits and vegetables reduces the risk of certain cancer diseases (Block et al., 1992; Williamson, 1996). Development of methods aimed at preserving the high level of antioxidants present at harvest during the postharvest handling will present new challenges for postharvest and food scientists.

It is hard to believe that wound ethylene is the only senescence-promoting factor in all harvested vegetables, especially in those harvested immature or in storage organs with a low metabolic rate. Vegetables harvested in a period of rapid growth, a common situation for broccoli, cauliflower and asparagus spears, constitute suitable plant material for investigations into harvest-mediated processes other than degradation of chlorophyll associated with senescence. Early responses of asparagus and broccoli tissues following harvest have been investigated to elucidate processes regulating harvest-induced senescence (Irving and Hurst, 1993; King and Morris, 1994). Within two to four hours after harvest, the tips of asparagus and broccoli florets lose large amounts of sucrose and undergo major changes in gene expression (King et al., 1995). These cellular responses lead to a markedly altered metabolism. Proteins and lipids are lost, free amino acids accumulate and a general loss of cellular compartmentation leading to tissue breakdown occurs. This seems to be a situation of starvation with progressive deprivation of sugars used as fuel for the rapid respiration. Such physiological changes are similar to starvation responses found in dark treated leaves (Peeters and van Laere, 1992). The termination of photosynthesis is accompanied by a disappearance of starch granules in the chloroplasts paralleling the rate of chlorophyll degradation.

Rapid physiological changes that occur immediately after harvest in immature vegetables require strategies to maintain quality during the postharvest chain. Rapid precooling immediately after harvest will reduce the metabolic rate and limit depletion of sugars and synthesis of undesirable compounds.

A common feature of most biennial vegetables is accumulation of mainly soluble carbohydrates in the storage organ during the first year. These carbohydrates are essential as fuel for respiration during winter dormancy and subsequent regrowth in the second season. Storage of biennial vegetables at 0–5°C for six months or more should not, due to the low metabolic activity, result in depletion of soluble carbohydrates that may promote senescence. After termination of dormancy and vernalization, the apex of cabbage and onions exerts an increasing sink strength with translocation of carbohydrates from the leaves to the core leading to senescence of the outer leaves with the senescence signal probably originating from the apex.

Fruit ripening can be considered as a presenescence phase since the fruit flesh will ultimately senesce with dispersal of the seeds. Ripening of fruits shows both similarities and differences compared with vegetative plant organs. Degradation of chlorophyll with or without synthesis of carotenoides is a part of the ripening process in some fruits while other fruits like kiwi and avocado remain green. Accumulation of free amino acids due to proteolysis, a common feature in leafy vegetables during senescence, is absent in fruits; instead synthesis of new proteins prevails. Changes in cell wall structure and composition occur during senescence of many plant organs but softening similar to that in fruit tissue is absent in leaves, stems and storage organs.

The physiological basis for fruit storage is closely associated with the stage of development at harvest and the rate of subsequent ripening and senescence. The eating quality of most fruit is dependent on proper timing and balance between the cascade of enzymatic processes each having its own temperature dependence. For most fruit the optimum ripening temperature is between 15 and 25°C independent of ripening behavior (climacteric-nonclimacteric) or sensitivity to chilling injury. On the other hand, some fruit store best at a temperature just above zero, some fruits should not be stored below 5°C, while those suffering from chilling injury have a temperature optimum ranging from 12 to 15°C. Generally, unripe fruits are more sensitive to chilling temperatures than ripe fruits.

Climacteric fruits are harvested preclimacteric with storage and handling methods aimed at inhibiting or delaying ripening until retailing. Although the synthesis of ethylene is on a low level in preclimacteric fruits, ethylene seems to play a crucial role during the onset of ripening. Too early harvest could result in an inability to ripen properly due to the absence of activated ethylene receptors (Lelièvre et al., 1997). Ethylene could also be used to promote fruit ripening (Reid, 1992).

Postharvest treatment of preclimacteric fruit with ethylene often contributes to a faster and more uniform ripening before distribution to the retailer. The role of ethylene during ripening of nonclimacteric fruit is still obscure. Ethylene treatment may promote ripening but the lack of autocatalytic ethylene production means that when moved to an ethylene-free atmosphere, ripening rate returns to the original level.

Some nonclimacteric fruit resemble vegetables (cucumbers, sweet pepper and squash) and are harvested and consumed unripe compared with most other fleshy fruits. Ripening, visualized by incipient degreening of greenhouse cucumbers, is regarded as a loss of quality and renders the fruits unsaleable compared with sweet pepper where a color change from green to red or yellow is not regarded as a loss of quality. Postharvest handling of these fruits therefore resembles that of leafy vegetables. Accumulation of soluble sugars during ripening and cell-wall softening, common features of most fleshy fruits, are absent in cucumbers and squash. Ripening of cucumbers is in addition to chlorophyll breakdown accompanied by an accumulation of citric acid in the endocarp tissue. The taste of a cucumber thus becomes more acid during the postharvest period.

Cell-wall structure and composition normally change during senescence of many plant organs. Fruit ripening differs clearly from most vegetative plant organs by softening of fruit flesh, which renders the fruit more palatable. Softening is a result of enzymatic fragmentation of pectic polymers in the middle lamella and hemicelluloses in the cell wall as well as degradation of starch and loss of turgor. These modifications reduce cell-to-cell adhesion and the strength of the cell walls, making the fruit more sensitive to bruising during postharvest handling (Miller, 1992). Control of tomato softening during postharvest storage allows harvest of vine ripe tomatoes with enhanced flavor. Such control has become an important tool for maintaining tissue integrity and tomato firmness during transport and marketing. Additional strategies include the introduction of 'Long life' cultivars, which are now grown as a winter crop in the Mediterranean for export to northern Europe as well as genetically manipulated fruits such as the 'Flavr Savr' tomato (Kramer and Redenbaugh, 1994).

Dormancy and Regrowth

Among those vegetables programmed for winter dormancy with subsequent regrowth and sprouting the following year, clear differences in the biennial behavior are evident. This is probably related to the climatic zone of their origin combined with breeding for adaptation to new growing areas. For cultivars suitable for the northern hemisphere, growth and development slow down and ultimately cease in the autumn due to decreasing day length, temperature and solar radiation leading to a stage of physiological maturity (Watada et al., 1984). Consequently, the rate of respiration at harvest is low compared with vegetables harvested during rapid growth. The storage potential of biennial vegetables is closely connected to the onset of dormancy in the autumn, which may either be imposed due to unfavorable growing conditions (carrot, beetroot) or hormonally controlled (onion, cabbage), and therefore a true dormancy or rest. The main part of most storage organs consists of parenchymatous tissue, and dormancy is therefore probably restricted to the apex and root primordia. The apex also holds a low temperature vernalization response with regrowth and sprouting after winter survival. Although the time span from harvest to sprouting may vary considerably according to cultivar and environment, most biennial vegetables are suitable for long-term storage (i.e., six months or more) providing the storage conditions suppress regrowth.

The few research reports available indicate that the postharvest physiology of biennials is controlled via a balance between the common identified phytohormones, i.e., auxins, gibberellins, cytokinins abscisic acid and ethylene (Isenberg et al., 1987). This makes correlative, quantitative hormone analyses tedious and difficult to interpret as an estimation of storability at harvest. During the harvest period the storage organ becomes an important sink for carbohydrates and nitrogenous compounds, which may be used as an indication of the progress of physiological maturity or dormancy. In cabbage cv. 'Hidena F_1' sucrose content in core and head leaves almost doubled over the harvest period in the autumn (Nilsson, 1988), but after mid-October sucrose accumulation leveled out. A possible reason for these pronounced changes in the carbohydrate composition is the cessation of differentiation of new leaves in the apical meristem supporting the assumption of a transition to dormancy.

Does the stage of physiological maturity at harvest have any influence on the storability of cabbage? During the first two months of storage at 2°C, the total amount of soluble sugars in cabbage cores decreased followed by a pronounced increase in sucrose content, indicating termination of the rest period. However, these changes in carbohydrate composition were only influenced by the time of harvest to a minor degree (Nilsson, 1993). The length of the dormant period during storage seems therefore not to be determined by the stage of physiological maturity at harvest provided the apex has entered the dormant period before the first harvest.

Compared with cabbage cores, the sucrose/hexose ratio in carrot roots increases over the harvest period but with minor changes in total soluble sugars. This behavior has been proposed as an indication of the degree of "physiological maturity" (Phan and Hsu, 1973; Fritz and Weichmann, 1979; and Nilsson, 1987). Carrots with a high sucrose/hexose ratio should therefore be more suitable for long-term storage. However, until present, no clear evidence has been reported to support this hypothesis. Probably, carrot roots do not enter to a period of true dormancy compared with onion bulbs and cabbage, which means that the composition of soluble carbohydrates should not indicate the termination of growth.

Onions are reported to enter into a true dormancy in the autumn (Abdalla and Mann, 1963). Hormonal changes inducing the dormant period are initiated while the onion is maturing in the field. In northern Europe, onions are harvested when the necks start bending and are left to dry in the field for about 10 days before being stored at low temperatures. Consequently, leaf senescence is advanced and the foliage dies back with a translocation of carbohydrates and free amino acids down to the bulb (Nilsson, 1980). Growth inhibitors synthesized in the leaves during termination of growth are, according to Stow (1976), translocated to the bulb and responsible for initiation and maintenance the bulb in a stage of dormancy. Too early harvest and leaf desiccation of the bulbs therefore may result in earlier sprouting due to too low a level of inhibitors in the bulb. Compared with cabbage, the carbohydrate composition of onion bulbs was equal whether harvested early (35% foliage of total plant weight) or late (2% foliage of total plant weight) (Nilsson, 1980).

The storage life of onions is mainly limited by sprouting. Breeding for delayed sprouting is one possible way (Miedema, 1994) since the use of artificial antisprouting compounds, e.g., maleic hydrazide, a chemical applied to the onion leaves when they are green and exporting photosynthates, will become banned in most countries. The use of ethereal oils for sprout suppression during storage of potatoes (Aliaga and Feldheim, 1985) seems more promising and should be investigated for vegetables as well.

PATHOLOGY

While the discussion in this chapter has focused on physiological changes during storage, an important problem related to long-term storage of vegetables is the incidence of various pathogens during storage. Postharvest infection often originates from inoculumn that has built up on the growing crop (Goodliffe and Heale, 1975). If the crop remains in the soil over the winter (e.g., storage of carrots by "strawing over"), the tissue will provide a substantial protection against pathogens. Harvest and subsequent storage normally damage the tissue and wound healing is often incomplete or delayed. The presence of various antifungal compounds in carrot with the highest concentration in the peripheral tissue at harvest (Olsson and Svensson, 1996) may explain why latent infections are progressing slowly during the first 3–4 months of storage. After this period of time, tissue resistance seems to decline and probably coincides with the termination of vernalization. Investigations into the nature of natural resistance against storage diseases therefore deserves further attention.

CONCLUSION AND FUTURE RESEARCH NEEDS

Research on the postharvest physiology and technology of fruit and vegetables in the last 20 years has led to many significant advances, with particular emphasis on improving appearance in the market place. Methods for produce handling and retailing are becoming more and more sophisticated. A challenge for the future will be how to integrate scientific results into commercial activities in the postharvest chain. While scientists are interested in determining the most suitable temperature or modified atmosphere for maintaining the shelf life of a certain product, distributors and retailers are often more interested in maintaining higher quality items within an acceptable shelf life.

Vegetables are sold according to their attractiveness and therefore must be exposed to the consumer in a way that normally does not allow proper temperature control. Possibilities for evaluating freshness, maturity and absence of blemishes before deciding on purchase are required by many consumers. Information about the origin of the produce, i.e., local production or import, as well as growing methods is also of increasing concern. Concomitantly, the enhanced use of trade-marks in vegetable marketing and consumer packaging is aimed at helping consumers to make the proper choice. Vegetables packed in a polymeric film and displayed in cool cabinets are therefore often considered as more suspect and less natural and environment-friendly than those displayed in open boxes. For supermarkets with a high daily turnover there should be no problem in fulfilling such needs but for small retailers this is hardly possible.

The growing interest in ready-to-use vegetables appeals to another segment of consumers willing to pay for both convenience and quality. The future for minimally processed, refrigerated vegetables packaged in a modified atmosphere depends on the ability of physiologists to deliver fruits and vegetables that are more exciting to consumers. Mixed packages of ready-to-use vegetables with improved flavor, color and texture might entice wary consumers to buy and eat vegetables they are less familiar with. Mixed vegetables present a new challenge to researchers who have mostly focused on a single vegetable. A mixed package means that the vegetables selected must be compatible within the anticipated storage conditions and duration.

Most of the efforts made to improve of quality during postharvest handling and storage of vegetables have focused on the interaction between the produce and the environment in order to obtain the most suitable conditions for maintaining freshness and shelf life. Other quality parameters such as flavor and texture deserve more attention in the future. When optimizing freshness is it possible to improve flavor without negative influences on the texture? Progress within molecular biology may make it possible to introduce dominant genes that confer ethylene insensibility to commodities that are sensitive to ethylene following harvest. Germplasm from stay-green mutants (Smart, 1994) could be used to avoid incipient yellowing of chlorophyll-containing vegetables. While consumers tend to focus on the extrinsic properties of fruit and vegetables, how much importance should be given to intrinsic quality characteristics by physiology? There is a need for easy and less expensive nondestructive methods for quality evaluation and control. An electronic nose is being tested for evaluation of flavor in food while infrared and fluorescence techniques show promise for quantifying different components.

Vegetables have an image as being healthy and nutritious. Postharvest physiologists must ensure that handling does not compromise nutrition at the expense of appearance. Between physiologists and the consumers lies the "Postharvest System" consisting of all the people involved: breeders, growers, packers, truck drivers, retailers, and so on— all devoted to offering the consumer the best of quality. It is obvious that further progress is not possible without an integrated view.

REFERENCES

Abdalla, A. A. and Mann, L. K. 1963. Bulb development in the onion (*Allium cepa* L.) and the effect of storage temperature on bulb rest. *Hilgardia* 35(5): 85–112.

Abeles, F. B., Morgan, P. W., and Saltveit Jr., M. E. 1992. *Ethylene in Plant Biology,* Academic Press, New York.

Ahvenainen, R. 1996. New approaches in improving the shelf life of minimally processed fruit and vegetables. *Trends in Food Science and Technology* 7: 179–187.

Aliaga, T. J. and Feldheim, W. 1985. Hemmung der Keimbildung bei gelagerten Kartoffeln duch das ätherische Öl der südamerikanischen Munapflanze (*Minthostachis* spp.). *Ernährung* 9: 254.

Amir-Shapira, D., Goldschmidt, E. E., and Altman, A. 1987. Chlorophyll catabolism in senescing plant tissue: *In vivo* breakdown intermediates suggest different degradative pathways for citrus fruit and parsley leaves. *Proc. Natl. Acad. Sci. USA* 84: 1901–1905.

ap Rees, T., Burrell, M. M., Entwistle, T. G., Hammond, J. B. W., Kirk, D., and Kruger, N. J. 1988. Effects of low temperatures on the respiratory metabolism of carbohydrates by plants, in *Plants and Temperature*, S. P. Long and F. I. Woodward, Eds., Society for Experimental Biology, London, pp. 377–393.

Arthey, D. and Dennis, C. (Ed.). 1991. *Vegetable Processing*, Blackie, and Son Ltd., London.

Barth, M. M., Perry, A. K., Schmidt, S. J., and Klein, B. P. 1992. Misting affects market quality and enzyme activity of broccoli during retail storage. *J. Food Sci.* 57: 954–957.

Biggs, M. S., Woodson, W. R., and Handa, A. K. 1988. Biochemical basis of high temperature inhibition of ethylene biosynthesis in ripening tomato fruits. *Physiol. Plant.* 72: 572–578.

Block, G., Patterson, B., and Subar, A. 1992. Fruit, vegetables and cancer prevention: A review of the epidemiological evidence. *Nutrition and Cancer* 18: 1–29.

Buchanan-Wollaston, V. 1997. The molecular biology of leaf senescence. *J. Exp. Bot.* 48: 181–189.

Burton, W. G. 1982. *Post-harvest Physiology of Food Crops*, Longman, London and New York.

Dhindsa, R. S., Plumb-Dhindsa, P., and Thorpe, T. A. 1981. Leaf senescence: Correlated with increased levels of membrane permeability and lipid peroxidation, and decreased levels of superoxide dismutase and catalase. *J. Exp. Bot.* 32: 93–101.

Elers, B. and Wiebe, H.-J. 1984a. Flower formation of Chinese cabbage. I. Response to vernalisation and photoperiod. *Sci. Hort.* 22: 219–231.

Elers, B. and Wiebe, H.-J. 1984b. Flower formation in Chinese cabbage. II. Antivernalization and short-day treatment. *Sci. Hort.* 22: 327–332.

Forney, C. F. 1995. Hot-water dips extend the shelf-life of fresh broccoli. *HortScience* 30(5): 1054–1074.

Fritz, D. and Weichmann, J. 1979. Influence of the harvesting date of carrots on quality and quality preservation. *Acta Hort.* 94: 91–100.

Goodliffe, J. P. and Heale, J. B. 1975. Incipient infections caused by *Botrytis cinerea* in carrots entering storage. *Ann. Appl. Biol.* 80: 243–246.

Heaton, J. W. and Marangoni, A. G. 1996. Chlorophyll degradation in processed and senescent plant tissues. *Trends in Food Science and Technology* 7: 8–15.

Heaton, J. W., Yada, R. Y., and Marangoni, A. G. 1996. Discoloration of coleslaw is caused by chlorophyll degradation. *J. Agric. Food Chem.* 44: 395–398.

Irving, D. E. and Hurst, P. L. 1993. Respiration, soluble carbohydrates and enzymes of carbohydrate metabolism in tips of harvested asparagus spears. *Plant Sci.* 94: 89–97.

Isenberg, F. M. R., Ludford, P. M., and Thomas, T. H. 1987. Hormonal alterations during the postharvest period, Ch. 12, in *Postharvest Physiology of Vegetables*, J. Weichmann, Ed., Marcel Dekker, Inc., New York, pp. 45–94.

Kadar, A. A., Zagory, D., and Kerbel, E. L. 1989. Modified atmosphere packaging of fruits and vegetables. *Crit. Rev. in Food Sci. and Nutr.* 28(1): 1–30.

King, G. A. and Morris, S. C. 1994. Early compositional changes during postharvest senescence of broccoli. *J. Amer. Soc. Hort. Sci.* 119(5): 1000–1005.

King, G. A., Davies, K. M., Stewart, R. J., and Borst, W. M. 1995. Similarities in gene expression during the postharvest-induced senescence of spears and natural foliar senescence of asparagus. *Plant Physiol.* 108: 125–128.

Klein, J. D. and Lurie, S. 1991. Postharvest heat treatment and fruit quality. *Postharvest News and Information* 2(1): 15–19.

Kramer, M. G. and Redenbaugh, K. 1994. Commercialization of a tomato with an antisense polygalacturonase gene: The FLAVR SAVER™ tomato story. *Euphytica* 79: 293–297.

Labuza, Th. P. 1982. Scientific evaluation of shelf life, Ch. 3, in *Shelf life Dating of Foods*, Th.P. Labuza, Ed., Food and Nutrition Press Inc., Westport, Connecticut, U.S., pp. 41–87.

Lelièvre, J.-M., Latché, A., Jones, B., Bouzayen, M., and Pech, J.-C. 1997. Ethylene and fruit ripening. *Physiol. Plant.* 101: 727–739.

Leshem, Y. Y. 1988. Plant senescence processes and free radicals. *Free Rad. Biol. and Medicine* 5: 39–49.

Miedema, P. 1994. Bulb dormancy in onion. I. The effects of temperature and cultivar on sprouting and rooting. *J. Hort. Sci.* 69(1): 29–39.

Miller, A. R. 1992. Physiology, biochemistry and detection of bruising (mechanical stress) in fruits and vegetables. *Postharvest News and Information* 3(3): 53N–58N.

Nanos, G. D., Romani, J., and Kadar, A. A. 1994. Respiratory metabolism of pear fruit and cultured cells exposed to hypoxic atmospheres: Associated change in activities of key enzymes. *J. Amer. Soc. Hort. Sci.* 119(2): 288–294.

Nilsson, T. 1980. The influence of the time of harvest on the chemical composition of onions. *Swedish J. Agric. Res.* 10: 77–88.

Nilsson, T. 1987. Carbohydrate composition during long-term storage of carrots as influenced by the time of harvest. *J. Hort. Sci.* 62(2): 191–203.

Nilsson, T. 1988. Growth and carbohydrate composition of winter white cabbage intended for long-term storage. I. Effects of late N-fertilization and time of harvest. *J. Hort. Sci.* 63(3): 419–429.

Nilsson, T. 1993. Influence of the time of harvest on keepability and carbohydrate composition during long-term storage of winter white cabbage. *J. Hort. Sci.* 68(1): 71–78.

Noodén, L. D., Guiamét, J. J., and John, I. 1997. Senescence mechanisms. *Physiol. Plant.* 101: 746–753.

Nussinovitch, A. and Lurie, S. 1995. Edible coatings for fruits and vegetables. *Postharvest News and Information* 6(4): 53–57.

Okada, K., Inoue, Y., Satoh, K., and Katoh, S. 1992. Effects of light on degradation of chlorophyll and proteins during senescence of detached rice leaves. *Plant Cell Physiol.* 33(8): 1183–1191.

Olsson, K. and Svensson, R. 1996. The influence of polyacetylenes on susceptibility of carrots to storage diseases. *J. Phytopathol.* 144: 441–447.

Paull, R. E. 1990. Postharvest heat treatments and fruit ripening. *Postharvest News and Information* 1(5): 355–363.

Peeters, K. M. U. and van Laere, A. J. 1992. Ammonium and amino acid metabolism in excised leaves of wheat (*Triticum aestivum*) senescing in the dark. *Physiol. Plant.* 84: 243–249.

Phan, C. T. and Hsu, H. 1973. Physical and chemical changes occurring in the carrot root during growth. *Can. J. Plant Sci.* 53: 629–634.

Philosoph-Hadas, S., Pesis, E., Meir, S., Reuveni, A., and Aharoni, N. 1989. Ethylene-enhanced senescence of leafy vegetables and fresh herbs. *Acta Hort.* 258: 37–45.

Philosoph-Hadas, S., Meir, S., Akiri, B., and Kanner, J. 1994. Oxidative defence systems in leaves of three edible herb species in relation to their senescence rates. *J. Agric. Food Chem.* 42: 2376–2381.

Prange, R. K. and Lidster, P. D. 1991. Controlled atmosphere and lightening effects on storage of winter cabbage. *Can. J. Plant Sci.* 71: 263–268.

Reid, M. 1992. Ethylene in postharvest technology, Ch. 13, in *Postharvest Technology of Horticultural Crops*, A. A. Kader, Ed., University of California, Division of Agriculture and Natural Resources. Publication 3311, pp. 97–108.

Rooke, E. A. and van den Berg, L. 1985. Equilibrium relative humidity of plant tissue. *Can. Inst. Food Sci. Technol. J.* 18(1): 85–88.

Saltveit Jr., M. E. 1989. A summary of requirements and recommendations for the controlled and modified atmosphere storage of harvested vegetables. *International Controlled Atmosphere Conference Fifth Proceedings, June 14–16 1989.* Wenatchee, Washington, USA. Vol. 2, Other Commodities and Storage Recommendations, pp. 329–352.

Saltveit Jr., M. E. and Morris, L. L. 1990. Overview on chilling injury of horticultural crops, Ch. 1, in *Chilling Injury of Horticultural Crops*, C. Y. Wang, Ed., CRC Press, Boca Raton, FL, pp. 2–15.

Sastry, S. K., Baird, C. D., and Buffington, D. E. 1978. Transpiration rates of certain fruits and vegetables. *ASHRAE Transactions* 84: 237–255.

Shewfelt, R. L. 1986. Postharvest treatments for extending the shelf life of fruits and vegetables. *Food Technol.* 40(5): 70–89.

Smart, C. 1994. Gene expression during leaf senescence. *New. Phytol.* 126: 419–448.

Stow, J. R. 1976. The effect of defoliation on storage potential of bulbs of the onion (*Allium cepa* L.) *Ann. Appl. Biol.* 84: 71–79.

Thorne, S. and Meffert, H. F. Th. 1979. The storage life of fruits and vegetables. *J. Food Quality* 2: 102–112.

Thorne, S. and Alvarez, J. S. S. 1982. The effect of irregular storage temperatures on firmness and surface colour in tomatoes. *J. Sci. Food Agric.* 33: 671–676.

Tian, M. S., Downs, C. G., Lill, R. E., and King, G. A. 1994. A role for ethylene in the yellowing of broccoli after harvest. *J. Amer. Soc. Hort. Sci.* 119(2): 276–281.

Tian, M. S., Woolf, A. B., Bowen, J. H., and Ferguson, I. B. 1996. Changes in color and chlorophyll fluorescence of broccoli florets following hot water treatment. *J. Amer. Soc. Hort. Sci.* 121(2): 310–313.

Tian, M. S., Islam, T., Stevenson, D. G., and Irving, D. E. 1997. Color, ethylene production, respiration and compositional changes in broccoli dipped in hot water. *J. Amer. Soc. Hort. Sci.* 122(1): 112–116.

Toivonen, P. M. A. and Sweeney, M. 1998. Differences in chlorophyll loss at 13°C for two broccoli (*Brassica oleracea* L.) cultivars associated with antioxidant enzyme activities. *J. Agric. Food Chem.* 46: 20–24.

Trebitsh, T., Goldschmidt, E. E., and Riov, J. 1993. Ethylene induces *de novo* synthesis of chlorophyllase, a chlorophyll degrading enzyme, in citrus fruit peel. *Proc. Natl. Acad. Sci. USA* 90: 9441–9445.

van Arsdel, W. B. 1957. The time-temperature tolerance of frozen foods. I. Introduction–The problem and the attack. *Food Technol.* 11: 28–33.

van Beek, G., Borrigter, H. A. M., and Nereng, B. 1985. The use of storability diagrams for mushrooms. Report No. 2299, Sprenger Instituut, Wageningen, NL.

Watada, A. E., Herner, R. C., Kadar, A. A., Romani, R. J., and Staby, G. L. 1984. Terminology for the description of developmental stages of horticultural crops. *HortScience* 19: 20–21.

Weichmann, J. 1977. Physiological responses of root crops to controlled atmospheres. *Proc. 2nd National Controlled Atmosphere Research Conf.* Mich. State Univ. Hort. Rept. 28: 122–136.

Wiley, R. C. (Ed.). 1994. *Minimally Processed Refrigerated Fruits and Vegetables,* Chapman and Hall, New York, London.

Williamson, G. 1996. Protective effects of fruit and vegetables in the diet. *Nutr. and Food Sci.* 1: 6–10.

Woods, J. L. 1990. Moisture loss from fruits and vegetables. *Postharvest News and Information* 1(3): 195–199.

Yamauchi, N. and Watada, A. E. 1991. Regulated chlorophyll degradation in spinach leaves during storage. *J. Amer. Soc. Hort. Sci.* 116: 58–62.

Yamauchi, N. and Watada, A. E. 1993. Pigment changes in parsley leaves during storage in controlled or ethylene containing atmosphere. *J. Food Sci.* 58(3): 616–618, 637.

Yamauchi, N. and Watada, A. E. 1998. Chlorophyll and xanthophyll changes in broccoli florets stored under elevated CO_2 or ethylene-containing atmosphere. *HortScience* 33: 114–117.

COMMON GROUND

- Quality of fresh fruits and vegetables is the result of a complex interaction of genetic, production and handling factors.
- Storage can only maintain, but not improve, product quality.
- Currently available quantitative measures of quality characteristics are difficult to relate to crop production or consumer preference.

DIVERGENCE

- The relative importance of genetic and cultural factors in determining quality at harvest.
- The feasibility of developing sufficient understanding of preharvest factors to predictably improve product quality.
- The ability to sufficiently understand consumer wants and needs to develop appropriate quality targets.

FUTURE DEVELOPMENTS

- A more comprehensive understanding of preharvest factors influencing quality at harvest and subsequent changes during postharvest handling.
- Markers of qualty that can be evaluated during both preharvest growth and postharvest storage.
- Models that more accurately assess preharvest influences on postharvest quality.

QUALITY

CONTEXT

- The ultimate judge of fruit and vegetable quality is the consumer.
- Product quality characteristics are readily measured by chemical and physical techniques.
- Consumer acceptability is difficult to measure because there is no uniformly defined consumer.

OBJECTIVES

- To present meaningful definitions of quality and acceptability in the context of fruit and vegetable distribution.
- To describe fruit and vegetable quality in terms of the consumer.
- To provide a better understanding of factors affecting consumer attitudes and behavior.
- To translate consumer theory into a practical means of measurement and evaluation.
- To present a formal mechanism that integrates measurement of fruit and vegetable characteristics with consumer expectations.

CHAPTER 7

Acceptability

LEOPOLD M. M. TIJSKENS

INTRODUCTION

QUALITY of products, and most certainly quality of perishable products, is very elusive and difficult to define. In almost any language, up to the Romans, the saying is known not to discuss color and taste. This means that every individual is entitled to judge the quality, and hence the acceptability, of any product according to his or her own standards however unclear or ill defined. And that is a good and fundamental right in any democratic state.

However, production and distribution of perishable products tailored to each individual preference can never be achieved in large-scale operations. So, the food industry has to develop standards of quality and of acceptability that cover the preferences of large groups of individuals. Studying these groups and establishing their preferences is the task of consumer and marketing research. Establishing general patterns in quality behavior and their description belongs to product research. Translating the established preferences and the quality behavior of perishable products simultaneously into a workable philosophy on consumers' view of quality belongs to nobody in particular but to everybody in general.

DECOMPOSITION OF QUALITY

Many studies have been undertaken to catch the elusive quality into a coherent theory, applicable in many situations. The objectives range

125

from consumer research (Meulenberg and van Trijp, 1996; Meulenberg and Broens, 1996; Wierenga, 1980), risk research (Frewer et al., 1993/ 94), market research (Steenkamp, 1987, 1989), product behavior (Kramer and Twigg, 1970) and quality control (Juran, 1974). Tijskens et al. (1994b) combined all these aspects into one theory on product quality, product acceptance and acceptability. This theory was further extended by Sloof et al. (1996), who used the principle of problem decomposition to arrive at a flexible but workable system, set up primarily for the purpose of modeling quality and acceptability behavior of perishable products. The main difference from previously developed systems is that the quality of a product is decomposed in assigned quality and acceptability as separate and distinct issues. The user/consumer assigns a quality to the product based on the intrinsic properties present in the product. Consequently, this assigned quality is more or less independent of applied criteria. The assigned quality is in fact the acceptance of a product (see next paragraph) without reference to any criterion. The sequence of steps taking place in the assignment of quality is schematically represented in Figure 7.1. The consumer, with his or her personal preference, strives for status and awareness of costs (among others), then judges the acceptability of a product based on this assigned quality with respect to a given situation like price and availability (see Figures 7.2 and 7.3).

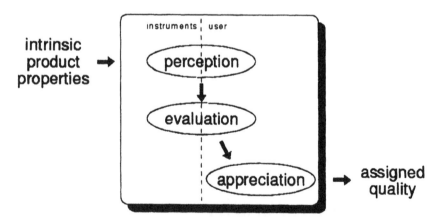

FIGURE 7.1 Quality is assigned to a product. The intrinsic properties are perceived and evaluated and an appreciation is determined.

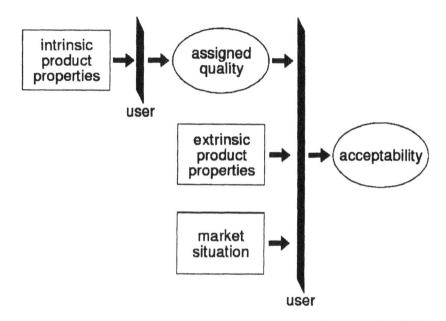

FIGURE 7.2 A user evaluates intrinsic properties to assign quality to a product. Taking extrinsic properties, his personal preferences and the market situation into account, acceptability is determined.

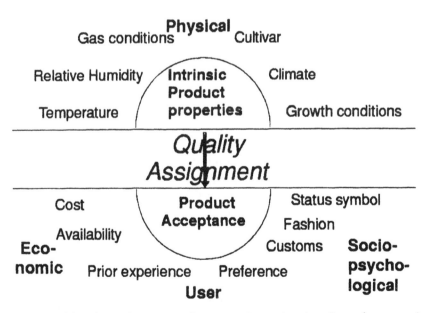

FIGURE 7.3 Relations between product properties, assigned quality and consumer's acceptance or appreciation.

ACCEPTANCE AND ACCEPTABILITY

A subtle but distinct difference exists in the linguistic meaning of the words "acceptance" and "acceptability." According to G. Hobson (personal communication), "acceptability suggests that some criterion is applied to differentiate those products that could be sold from those that could not, whereas acceptance does not imply any aspect involving a range in quality." In other words, acceptance describes a general state of a product, whereas acceptability describes whether a particular product is accepted by a consumer in his or her particular circumstances, e.g., status, price, personal preferences, as can be taken from the graphical representation (Figures 7.2 and 7.3).

DECOMPOSITION OF ACCEPTABILITY

So, when we are talking about acceptability, some criterion has to be set to judge the general acceptance or assigned quality of a product. As such, acceptability covers both product aspects (product research) and consumer aspects (consumer and market research) at the same time. To be properly applicable in practice, both types of aspects should be and can be studied separately without (too much) interference of one another.

Throughout this chapter this criterion is called "quality limit." It stands for the criterion a consumer applies to the quality aspect involved to determine the acceptability of a given product.

Consumer Applied Criteria

The criteria a particular consumer applies to the assigned quality in determining the product's acceptability are frequently adapted to the situation of the individual. To obtain a more general view on acceptability criteria, consumer research has to aim at particular groups of consumers with roughly the same applied criteria in the same situation. Based on these groups of similar consumers the market strategies of companies selling commodities are defined. That is, advertisement tries to affect the criteria applied by that group of consumers. Price policies try to lower the consumer's financial threshold. Hence, both efforts affect the acceptability of the product in a different way. All these actions, however, alter neither the intrinsic properties nor the assigned quality. Consequently, they have no effect on the description of product behavior during growth (production), storage and processing. This signifies that the same dynamic product model can be used over and over for the description and prediction of the intrinsic product quality at all conditions in the integrated view of fruit and vegetable quality.

Assigned Quality

The quality of a product and its dynamic behavior during storage, transport and distribution can be based on the behavior of all the intrinsic properties of a product that have a bearing on the quality perception of the consumer or a specific group of consumers. If we know which properties are used to arrive at an assigned quality (determination) and how they are used (relative importance), quality behavior may be modeled by modeling the appropriate properties. This information is known as the "quality function" (Tijskens et al., 1994b; Sloof et al., 1996). Many models have been developed, consciously or unconsciously, based on this philosophy (Tijskens and Evelo, 1994c; Tijskens et al., 1994a, 1996a, 1996b; Verlinden, 1996).

EXAMPLES OF MODELING ACCEPTABILITY

When describing or modeling acceptability of (perishable) products, one has to realize that consumers can assign quality to each product item separately, e.g., the taste of one tomato. In acceptability, however, the quality aspect is combined with the economic aspect. Consumers may buy one individual tomato, but commercial applications inherently cover complete batches of products. As such, within the aspect of acceptability a second aspect comes into play: How many individual items in a batch may be unacceptable before the batch as a whole becomes unacceptable? This aspect introduces a system of population dynamics in the description of batch behavior.

Acceptability of Potted Plants

The behavior of acceptability of batches of products has been modeled for potted plants during storage in darkness (Tijskens et al., 1996b). In the assessments of the potted plants, the applied quality criteria and the quality limits were predefined and related among others to number of leaves and flower without defects. The limit of acceptability has therefore a fixed value and is consequently not expressed explicitly in the equations used. As long as the selected quality criteria are applied consistently throughout the experiments, neither the type nor the value of the criteria for acceptability is important for the development of the model. Acceptability is subject to local and regional preferences; hence, to ensure generic application, the model has to be independent of acceptability criteria. If the criteria of acceptability do not change during the assessment period, the decrease in acceptability is completely determined by the loss of those intrinsic product properties that contribute to the perceived quality.

The plants were given a standardized pretreatment, stored in darkness at six constant temperatures. Up to 21 days, quality was evaluated six times. The experimental conditions of time and temperature were chosen to simulate the conditions encountered during container transport to medium and distant markets.

Time Effects

Loss of perceived quality over time is undoubtedly the result of a cascade of biochemical reactions. Such cascades can often be approximated by the sigmoid logistic curve. A logistic function is therefore likely to be a good model. The formulation for the logistic curve has been slightly adapted to enable the introduction of the effect of temperature in a way suitable for the dynamic approach:

$$N = \frac{N_{max}}{1 + \left(\dfrac{N_{max}}{N_0} - 1\right)e^{kt}} \tag{7.1}$$

where N_{max} is the number of plants in the batch, k is the rate of decrease in acceptability, t is the time in days, N is the number of acceptable plants, N_0 is the initial number of acceptable plants. $[(N_{max}/N_0) - 1]$ constitutes a correction for the biological age of the plants at the start of the experiment (Thai et al., 1990; Tijskens and Evelo, 1994c) and is independent of the temperature applied in the subsequent treatment.

Temperature Effects

The rate k in Equation (7.1) represents some combination of all biochemical reaction rates in a cascade of reactions. The rate k will therefore depend on temperature. The effect of temperature on the behavior of potted plants can be twofold, with both high and low temperatures leading to a faster reduction in acceptability (see Figures 7.4 and 7.5). This suggests two processes, one that is particularly active at high temperatures and results in, for example, excessive water loss, and one that

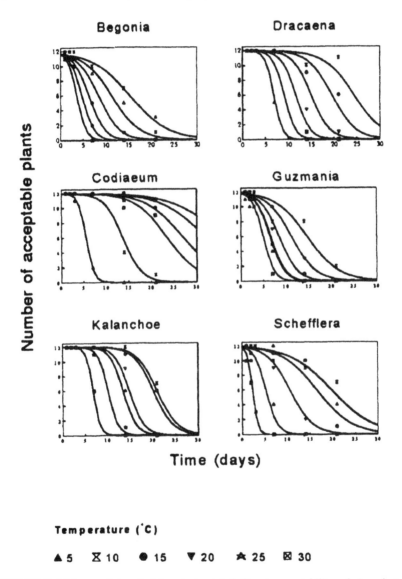

Time (days)

Temperature (˚C)

▲ 5   ✕ 10   ● 15   ▼ 20   ✸ 25   ⊠ 30

FIGURE 7.4 Measured (symbols) and simulated (lines) acceptability of six selected potted plant species versus storage time for the six experimental storage temperatures.

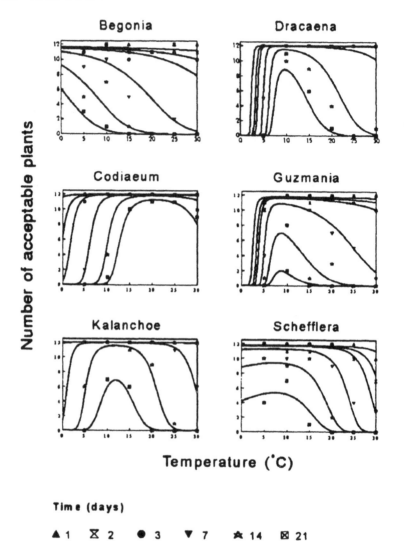

Temperature (°C)

Time (days)

▲ 1   ⊠ 2   ● 3   ▼ 7   ✶ 14   ⊠ 21

FIGURE 7.5 Measured (symbols) and simulated (lines) acceptability of six selected potted plant species versus temperature for the six experimental storage times.

is particularly active at low temperatures and results in, for example, membrane deterioration and chilling injury (Tijskens et al., 1994a). The acceptability will decrease with an apparent rate equal to the sum of these reaction rates:

$$k = k_c + k_h \qquad (7.2)$$

with k_c is rate constant of the chilling process, and k_h is rate constant of the heat process. Both rate constants k_c and k_h will, as for all chemical reactions, depend on temperature according to Arrhenius' law:

$$k_i = k_{i,\text{ref}}\, e^{\frac{E_i}{R}\left(\frac{1}{T_{\text{ref}}} - \frac{1}{T_{\text{abs}}}\right)} \qquad (7.3)$$

where k_i is the rate constant of process i (c = chilling injury, h = high temperature deterioration), $k_{i,\text{ref}}$ is the rate constant of process i at reference temperature T_{ref} (15°C), E_i is the energy of activation of that process, and T_{abs} is the actual absolute temperature in the experiment.

Results

With these equations, the data of 20 species potted plants were analyzed using multiple nonlinear regression, and the parameters estimated. In Figure 7.4 the acceptability is shown for six selected species as a function of storage time, Figure 7.5 shows the acceptability of the same six species as function of temperature. Except for two species, the explained variance (R^2_{adj}) exceeded 90%, and was more than 95% in 8 of the 20 species. The kinetic parameters of the model can directly be interpreted in terms of chilling and/or heat sensitivity. Fifteen species have a distinct value for the reference rate for the chilling process (k_{cref}) while five species have no distinct value for that parameter. These results are in agreement with the generally accepted behavior for these species.

From Figure 7.5 it is clear that the duration of storage in darkness has a pronounced effect on the apparent rate of decrease in acceptability: The longer a species has been stored or transported in darkness, the narrower the optimal temperature region becomes, the faster the loss in acceptability becomes, assuming some plants are still saleable, and the more pronounced differences due to temperature become. A practical consequence is that more care should be taken in selecting the optimal transport condition for transportation of longer duration.

From the model formulation it follows how the initial condition of the potted plants (N_0) will affect the maximal acceptable transportation period. N_0 also reflects the differences obtained in experiments (e.g., replications) with batches with apparently identical starting quality. The more N_0 deviates from the maximum value N_{max}, the sooner the plants will become unacceptable at any temperature. Figure 7.6 gives an example for the species *Dieffenbachia,* stored at 20°C with an initial number of acceptable potted plants ranging from 11.94 to 11.999 on a total

Acceptability (number)

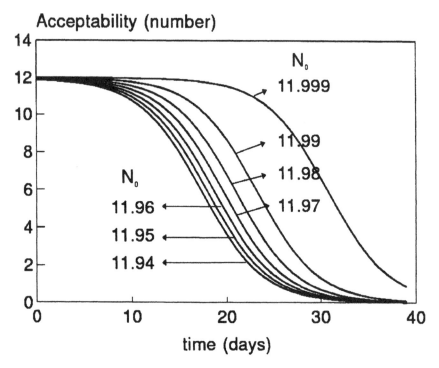

FIGURE 7.6 The effect of initial condition (vitality) at 20°C for species *Dieffenbachia*.

number of 12 plants in each batch. As can be seen, a small difference in initial vitality (N_0) has a marked effect on the change in time of the acceptability when near the maximal possible count (12), and a far lesser effect at lower conditions of initial vitality.

The dynamic model allows calculation of the effects of storage in darkness on the acceptability as a function of variable temperature during storage. This provides a tool that allows wholesalers of potted plants and transport companies to calculate how long and within what temperature limits a product can be transported without adversely affecting its acceptability. It also allows insight into the optimum temperature and time for simultaneous transportation of several species of potted plants, with the quality of the most sensitive product as the limiting factor.

Keeping Quality of Perishable Produce

The effect of changing levels of quality criteria applied by consumers or users cannot be taken from the type of model as used in the potted plant example, as it is based on data gathered from experiments using exclusively fixed acceptability limits. To develop models that include variable quality criteria, we have to search deeper into the mechanisms involved in quality decay. In the model on keeping quality (Tijskens, 1995; Tijskens and Polderdijk, 1994d, 1996a), these aspects are worked out for the four most commonly occurring mechanisms. All four mechanisms resulted in the same generic formulation of keeping quality show in Equation (7.4):

$$KQ = \frac{KQ_{ref}}{\sum\limits_{i=1}^{N} k_i} \tag{7.4}$$

where KQ is the keeping quality, k is the rate constant for each of the occurring quality decaying processes i and KQ_{ref} is the keeping quality at reference temperature. Keeping quality may be defined as the *time* the product's quality *remains acceptable* during storage or transport. Shelf life may be defined as the keeping quality at standardized conditions. The definition of acceptability, that is, which product properties are included and which quality limits are applied, remains rather undefined and changes from produce to produce, from consumer to consumer and from situation to situation.

Effect of Quality Limits

To be flexible and generic, the quality limit, applied by a consumer or group of consumers, has to be included in the model formulation. Based on an assumed exponential mechanism as an example for the quality attribute(s) involved, the model on keeping quality is explained and the effects of different limits and initial qualities are shown. Other mechanisms are possible, with a different relation for the limit applied.

$$KQ = \frac{\ln\left(\dfrac{Q_0}{Q_l}\right)}{\sum\limits_{i=1}^{N} k_i} \tag{7.5}$$

where Q is the quality of the produce, 0 at initial time, and l the limit value. The relation deduced for keeping quality of perishable products is shown in Equation (7.5). We see that (for this exponential mechanism of quality decay) the effect of different quality limits on the obtained keeping quality is a logarithmic one. This signifies that the change in keeping quality for a change in quality limit Q_l is lower at higher limits. It also signifies that the same effect can be obtained by increasing the initial quality by, e.g., 10%, as by decreasing the quality limit by 10% and with increasing the confidence of the consumer in your product instead of losing it.

In Figure 7.7 an example is given of the effect of (nationally) changing quality limits during international transport of cauliflower. Each time a border is crossed, the quality limit changes to lower a level with an apparent increase in remaining keeping quality as a consequence. The quality of the product itself is not affected by this process. It takes only longer to reach the newly applied quality limit.

Temperature Effects

The effect of temperature can again, as for the model on potted plants, be (at least) twofold ($N = 2$ in Equation (7.5)). Most products of moderate origin are only susceptible to deterioration at higher temperatures. However, most products of tropical or subtropical origin are also highly

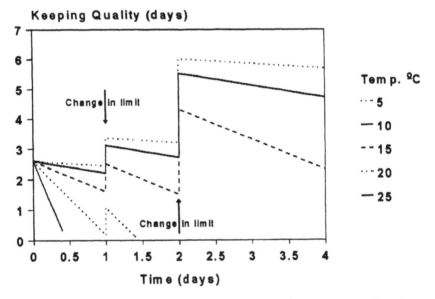

FIGURE 7.7 Effect of discrete changes in quality limits on the keeping quality of cauliflower at five constant temperatures (exponential decay).

sensitive to deterioration at low temperatures (chilling injury). This difference in behavior is reflected in the equation by the sum of reaction rate constants in the denominator where N stands for the number of processes involved (or the number of attributes that can be limiting). These reaction rate constants again depend on temperature according to Arrhenius' law (Equation (7.3)).

In Figure 7.8 an example is given for the behavior of keeping quality for paprika (bell pepper) at constant storage conditions. The two temperature-driven processes can clearly be seen. A discrete change in quality limit constitutes an instantaneous change of the numerator in Equation (7.5), but still depends on temperature by the sum of reaction rate constants in the denominator. Temperature not only has an effect on keeping quality itself, changes in temperature also affect the importance of a change in quality limit. From Figure 7.7 it can be seen that the effect on keeping quality exerted by a changing quality limit is different for each temperature. The slower the processes of quality decay

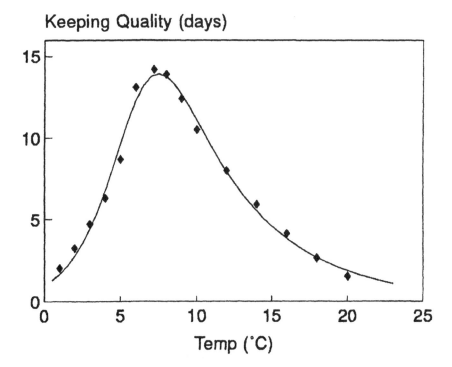

FIGURE 7.8 Keeping quality of bell peppers as a function of constant storage temperature.

are, either at low temperature (no chilling injury, see 5 and 10°C in Figure 7.7) or near the optimal temperature (chilling injury), the larger the effect of temperature becomes, due to the sum of rates in the denominator of Equation (7.5). This effect partly explains the difficulties encountered in practical conversion of keeping quality rules of thumb from one country to another.

Effects of Initial Conditions and Harvest Time

Lange and Cameron (1994) studied the effect of diurnal harvest time on the keeping quality of sweet basil. Analysis of the data revealed that the diurnal harvest time only affects the initial quality Q_0, without affecting the reaction rate constants. The initial quality could be described by a sinusoidal function, linked to the rise and fall of the sun during a day's period. Increasing the initial quality greatly increased the keeping quality and consequently the acceptability of the product. Again, as for the potted plants, acceptability can be increased by increasing the initial intrinsic quality of the product without losing the consumer's confidence in the product by playing too close to the acceptability limit.

Variance in Batches

Every batch of products consists of individuals. Despite the apparent identical stage of maturity, (external) quality and appearance, all these individuals are actually at different stages of maturity. The number of individuals in a batch that have to be acceptable for the entire batch to be acceptable (batch quality limit) depends on the preferences of the user(s) and the intended application. The first individuals in a batch that become unacceptable determine the acceptability of that batch. Inherently that is those individuals that are more developed/mature to begin with. This fact puts a large emphasis on the type of distribution of the quality attribute in a batch of products with apparently an identical state of maturity.

With the recently developed very accurate techniques of quality measurements, especially computer imaging techniques for measuring color and color models similar to the model for the potted plants (see Equation [7.1]), it has become within reach to study the distribution of color in a batch of products. Schouten and van Kooten (1998) reported on the type and change of the distribution of color in batches of cucumbers. They found that the distribution of color over the individuals in a batch is skewed when the cucumbers are picked in a very early stage of maturity. That is, the human eye cannot discriminate between the color of the individuals: They all are equally dark green. Upon maturation the distribution becomes less and less skewed, but still dark green, and approaches the normal Gaussian distribution. The amount of skewness, that is, how many cucumbers seem to be outliers with respect to the Gaussian distribution, depends, among other factors, on the growing conditions.

Figure 7.9 gives an example for batches of cucumbers grown in two different plant densities and with two levels of nutrients. It is evident from this figure, that the skewness of the distribution has a marked effect on the development of the batch rated as acceptable. For batches with equal mean color value, the batch with the largest skewness becomes unacceptable first.

How the distribution changes upon storage and further maturation can be seen from Figure 7.10 where the background color of batches of apples was measured individually during the ripening at several constant temperatures (unpublished, data M. Simi, University of Ljubljana,

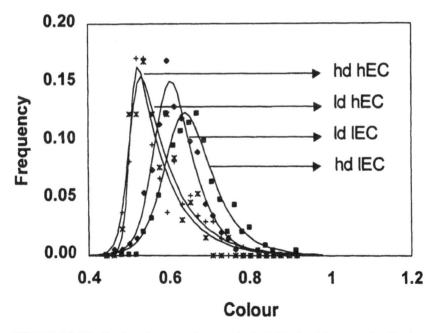

FIGURE 7.9 Distribution of green color over the individuals of the cucumber batches for two levels of plant density (ld and hd) and two levels of nutrients (lEC and hEC) during growing. (Courtesy ISHS.)

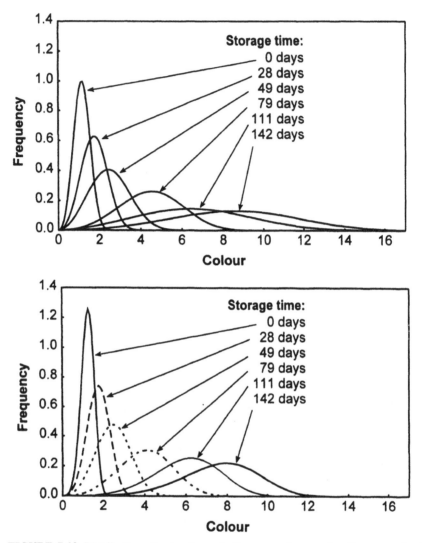

FIGURE 7.10 Distribution of color (green/red) measured as a-value for one orchard harvested at early (top) and mature (bottom) stages of maturity and stored during 142 at 10°C.

Slovenia; analysis P. Konopacki, Research Institute of Pomology and Floriculture, Skierniewice, Poland). It can be seen how the estimated distribution changes with storage time, and that they are somewhat different for apples of two successive harvest dates from the same orchard. What also can be seen from Figure 7.10 is that the apparent rate of color development for more mature individuals is faster than for the mean of the batch, with in turn is faster than the less mature individuals in the same batch. The same difference in apparent rate of maturation of batches of perishable products was reported by Tijskens and Wilkinson (1996b).

So, in conclusion, the mean value of different batches can be the same, but the skewness of the distribution and the rate of development will determine how soon a given batch will become unacceptable.

CONCLUSIONS

The fact that the batch acceptability of all the studied potted plants and that the keeping quality of all fruits and vegetables comply with the same models enhances their validity. It implies that a generic approach (one model for all potted plants tested, one model for keeping quality) is valid. The same increase in acceptability and in keeping quality can be obtained by increasing the initial quality of the product as by decreasing quality limit, however, without endangering the confidence of the consumer.

It is possible to combine the behavior of intrinsic product properties, varying initial quality, variable quality limits and consumer behavior into mathematical models. These models greatly enhance our understanding of the complex combined behavior of acceptability. Problem decomposition is a very valuable tool for developing mathematical models and descriptions of complex behavior in living materials. This makes it even possible to describe mathematically the effect of rather ill- or undefined entities like personally and regionally applied quality limits. The skewness of the distribution of quality attributes, in batches of products with the same average quality level, largely determines the keeping quality of that particular batch.

HOW TO DESIGN FURTHER RESEARCH

To get a better understanding of acceptability and keeping quality of perishable products, and a better feeling for the effects of differences between apparently identical batches, future research should be designed along a new and extended road.

- First of all, the relations depicted between quality attribute, acceptability and keeping quality need more attention and validation to ensure their application in a generic fashion. Can we really rely on intrinsic properties as sole descriptors of product quality during the entire food chain? Is it really possible to decompose the quality problem into a problem of intrinsic and extrinsic properties and attributes?

- In view of the importance of the initial conditions in a batch, especially when the level of those initial conditions is very low and small differences have potentially a major effect, new and very sensitive techniques should be developed to measure the appropriate quality attributes and quality properties. For color of all kinds of fruits and vegetables, this technique is already available in the computer image analysis system. For quality attributes like firmness and amount of bacterial infection, more sensitive and nondestructive techniques have to be developed.
- The effect of distributions of a quality attribute or product property in a batch needs to be studied in more detail, and the possibilities for practical applications have to be assessed.
- The design of research experiments should focus on nondestructive techniques that make it possible to reuse and reassess the same individual fruit at any later time. Individual monitoring and its administration should receive much more attention in practical research to fully use its potential for chain management and optimization.

REFERENCES

Frewer, L. J., Raats, M. M., and Shepherd, R. 1993/94. Modelling the media: the transmission of risk information in the British quality press. *IMA Journal of Mathematics Applied in Business & Industry*, 5, 235–247.

Juran, J. M. 1974. Basic concepts, in *Quality Control Handbook*. 3rd ed, Juran, J. M., Gryna, F. M., Jr. Bingham, R. S. Jr., eds., McGraw-Hill, New York.

Kramer, A. and Twigg, B. A. 1970. *Quality Control for the Food Industry*, AVI Pub. Co., Westport, CT.

Lange, D. D. and Cameron, A. C. 1994. Postharvest shelf life of sweet basil (*Ocimum basilicum*). *HortScience*, 29, 102–103.

Meulenberg, M. T. G. and van Trijp, J. C. M. 1996. Winkelkeuze gedrag van consumenten (in dutch. Selection behavior of consumers in shops). *MAB*, 70, 191–198.

Meulenberg, M. T. G. and Broens, D. F. 1996. Trends in de productie van voedingsmiddelen. Inzicht in ketens helpt bij ketenvorming (in dutch. Trends in food production. Understanding distribution chains helps building distribution chains). *Voedingsmiddelen technologie*, 29, 12, 19–25.

Schouten, R. E. and van Kooten, O. 1998. Keeping quality of cucumbers: is it predictable? in Tijskens, L. M. M., Hertog, M. L. A. T. M. (eds.), *Proceedings International Symposium on Applications of Modelling as an Innovative technology in the Agri-Food-Chain Model-IT: Acta Horticulturae 476*, 349–355, ISHS, Leiden, NL.

Sloof, M., Tijskens, L. M. M., and Wilkinson, E. C. 1996. Concepts for modelling the quality of perishable products. *Trends in Food Science & Technology*, 7, 165–171.

Steenkamp, J. B. E. M. 1987. Perceived quality of food products and its relationship to consumer preferences: Theory and measurement. *Journal of Food Quality*, 9, 373–386.

Steenkamp, J. B. E. M. 1989. Product quality: an investigation into the concept and how it is perceived by consumers. Ph.D. Thesis, Agricultural University Wageningen, The Netherlands.

Thai, C. N., Shewfelt, R. L., and Garner, J. C. 1990. Tomato color changes under constant and variable storage temperatures: Empirical models. *Trans. Amer. Soc. Agric. Eng.*, 33, 607–614.

Tijskens, L. M. M., Otma, E. C., and van Kooten, O. 1994a. Photosystem 2 quantum yield as a measure of radical scavengers in chilling injury in cucumber fruits and bell peppers. *Planta*, 194, 478–486.

Tijskens, L. M. M., Sloof, M., and Wilkinson, E. C. 1994b. Quality of perishable produce. A philosophical approach. Final COST94 seminar, October, Oosterbeek, The Netherlands (in press).

Tijskens, L. M. M. and Evelo, R. 1994c. Modelling color of tomatoes during postharvest storage. *Postharvest Biology & Technology*, 4, 85–98.

Tijskens, L. M. M. and Polderdijk, J. J. 1994d. Keeping quality and temperature, quality limits and initial quality. *Proceedings COST94 Workshop Quality Criteria*, pp. 15–23. April 12–19, 1994, Bled, Slovenia.

Tijskens, L. M. M. 1995. A generic model on keeping quality of horticultural products, including influences of temperature, initial quality and quality acceptance limits. *Proceedings 19th International Congress of Refrigeration (IIR)*, Vol II, pp. 361–368, August 1995, The Hague, The Netherlands.

Tijskens, L. M. M. and Polderdijk, J. J. 1996a. A generic model for keeping quality of vegetable produce during storage and distribution. *Agricultural Systems*, 51, 4, 431–452.

Tijskens, L. M. M., Sloof, M., Wilkinson, E. C., and van Doorn, W. G. 1996b. A model of the effects of temperature and time on the acceptability of potted plants stored in darkness. *Postharvest Biology & Technology*, 8, 293–305.

Tijskens, L. M. M. and Wilkinson, E. C. 1996c. Behavior of biological variabi slity in batches during post-harvest storage. *AAB Modelling Conference: Modelling in Applied Biology—Spatial Aspects*, pp. 267–268, 24–27 June, Brunel University, Uxbridge, U.K.

Verlinden, B. E. 1996. Modelling of texture kinetics during thermal processing of vegetative tissue. Ph.D. thesis, October 1996, Kath. Univ. Leuven, Belgium.

Wierenga, B. 1980. Multidimensional models for the analysis of consumers' perception and preferences with respect to agricultural and food products. *Journal of Agricultural Economics*, 31(1), 83–97.

CHAPTER 8

Fruit and Vegetable Quality

ROBERT L. SHEWFELT

INTRODUCTION

THE commercial approach to food quality is undergoing both an evolution and a revolution. Quality evaluation systems are evolving from a
perspective of Quality Control to Quality Assurance to Quality Management. These changes have necessitated revolutionary switches of emphasis from inspecting quality to improving processes that result in
enhanced quality. They also assume a shift from a product orientation
to a consumer orientation in defining quality. Rather than providing intellectual leadership in a reshaping of thought on quality as it applies to
processed products, academic research in food science has generally regarded quality as a tool to evaluate other objectives (e.g., food process
development, food product development, new package assessment).
Thus, food quality is still generally defined by the investigator in terms
of clearly measurable characteristics rather than in terms of consumer
acceptability. Within a food process, tradeoffs frequently occur such that
quality can be optimized and not maximized. Thus, changes in quality
of a specific product undergoing a specific process are described in relative changes of specific characteristics (e.g., color, flavor or texture)
rather than a composite of these characteristics.

Some notable similarities and differences exist in the approach to
quality by postharvest physiologists and food scientists. Like the food
scientist, the postharvest physiologist views quality from a product ori-

144

entation. Both types of investigators note major changes in quality degradation during the food process or postharvest storage. Thus, both assume that quality cannot be improved during processing or storage, merely maintained. However, the postharvest physiologist differs in perspective from the food scientist in that the former thinks of quality in terms of maximization while the latter thinks in terms of optimization. The quality ideal of a specific fruit or vegetable is the premium cultivar, grown under optimal conditions, harvested at peak maturity and consumed within minutes or hours after harvest. Although quality is measured in terms of specific characteristics (e.g., color, flavor or texture), it is generally described more in terms of acceptability (e.g., superior, good, fair, poor).

In an attempt to develop ways of defining fruit and vegetable quality in terms of the consumer, it has been necessary to challenge some of the widely held assumptions of quality both by food scientists and postharvest physiologists (Shewfelt, 1999). The Quality Enhancement (QE) model (see Figure 8.1) has been proposed to redefine quality in terms of the consumer and thus lead to improved quality of fresh fruits and vegetables (Shewfelt, 1994; Malundo et al., 1997; Shewfelt et al., 1997). This chapter highlights the major principles that form the basis for the model and provides two examples of how the model has been applied.

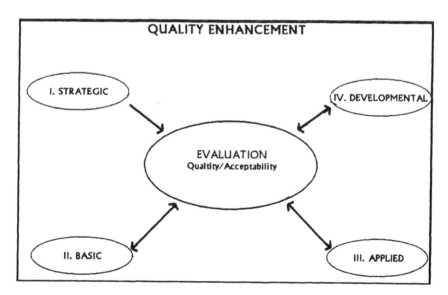

FIGURE 8.1 Quality enhancement model. Reprinted with permission from Shewfelt et al., 1997, *Food Technology*, 51(2): 56–59.

QUALITY PRINCIPLES

One of the limiting factors in postharvest research is the lack of understanding of consumer behavior. Many uncertainties about accurate measurement of consumer response and the ability to apply such knowledge to commercial situations have limited study in the area. Progress in this area will require approaches and specific methods to determine: (1) the *relevance of* measured *quality* characteristics *to* actual *consumer acceptability;* (2) a *common understanding of terms* like food quality, shelf stability and consumer acceptability among chemists, engineers, postharvest physiologists and sensory specialists; (3) *quantitative assessment of consumer acceptability,* (4) under what *conditions consumers are valid assessors* of quality and acceptability and when are they not; (5) types of *assessment tools* needed *to differentiate between defect and excellence* orientations of quality characteristics; and (6) how *factors* that affect consumer acceptability that are *intrinsic to the product* can be *separated from those* that are *intrinsic to* individual *consumers.* Several principles described in the scientific literature provide clues to new approaches, but little work in these areas is currently being conducted.

The *relevance of quality to consumer acceptability* as quality is normally measured has been questioned (Pendalwar, 1989). Selection of quality criteria for evaluation is frequently based on the accuracy and precision of the results rather than on relevance to acceptability in the marketplace (Shewfelt and Phillips, 1996). Too often quality and acceptability become merely buzzwords that have no real meaning within the scientific community. Measurements of quality characteristics represent an emphasis on "internal validity" (accuracy and precision of the quality characteristics measured) of the research data (van Trijp and Schifferstein, 1995), but acceptability requires "external validity" (ability to predict performance of a product in the marketplace) to be commercially significant. To establish true relationships between quality and acceptability studies must merge the concepts of internal and external validity.

A *common understanding of terms* like food quality, shelf stability and consumer acceptability among chemists, engineers, postharvest physiologists and sensory specialists would permit a more concerted research effort on improving fresh fruit and vegetable quality. A major difficulty in understanding quality is a lack of a common language. Numerous definitions exist for quality (Kramer and Twigg, 1970; Pirsig,

1974; Crosby, 1979; Juran and Gryna, 1980; Deming, 1986; Surak and McAnelly, 1992; Bounds et al., 1994), but few for shelf life (Shewfelt, 1986), or acceptability (Land, 1988). In general, chemists tend to view quality as a function of qualitative and quantitative differences of individual compounds present in a food product (Fennema, 1996), engineers as a bundle of characteristics that can be measured with accuracy and precision (Taoukis and Labuza, 1996; Hendrickx et al., 1994), postharvest physiologists as a condition to be attained (Kays, 1991), and sensory specialists as attributes that can be described by trained panels (Lyon et al., 1993). Most of these interpretations view quality as properties intrinsic to the product itself. Adoption of a unified language for these three terms is essential in comparing subsequent studies.

Quantitative assessment of consumer acceptability is necessary for the adoption of a consumer-oriented approach to quality improvement. None of the groups of scientists listed above has been actively involved in developing new means of assessing acceptability. All have been more concerned with the internal validity of their research than external validity. Quality is viewed as a property that decreases over time, which at best can be maintained but never improved.

Chemists trace changes of individual compounds such as pigments, sugars, acids, flavor volatiles (Shewfelt, 1994) or vitamins (Clydesdale et al., 1991). Biochemists trace changes in enzyme activity or gene expression (Tucker and Grierson, 1987). Engineers and postharvest physiologists trace changes in quality characteristics such as flavor, color, texture, or loss in vitamins (Kays, 1991), with engineers particularly concerned with accuracy and precision of results and kinetic modeling of the data (Taoukis and Labuza, 1996). Sensory specialists measure changes in descriptive notes (Lyon et al., 1993; Lawless and Heymann, 1998). Consumer panels are used to assess overall acceptability (Moskowitz, 1994; Lawless and Heymann, 1998). Scales are employed to express acceptance on a Hedonic scale, frequently 9 points (Peryam and Pilgrim, 1957) or on a 100-point basis (Moskowitz, 1994), although some studies tend to use descriptive panels to assess this trait with dubious validity.

Chemists, engineers and postharvest physiologists tend to shy away from any contact with consumers, with the data considered completely unreliable. A literal interpretation of Land's (1988) definition of acceptability ("the level of continued purchase or consumption by a specified population," p. 476) suggests that measurement is not possible outside of the market. Obviously, some premarket predictor of acceptability is needed to be useful in assessing quality. A potential answer lies in expressing acceptability as a probability distribution of percentage of a population finding the product acceptable (Walters and Bergiel, 1989; Shewfelt et al., 1997) using either a 5-point purchase intent scale (Moskowitz, 1994), a 3-point acceptability scale (Shewfelt et al., 1997), or even an accept-reject forced choice.

The *conditions* under which *consumers are valid assessors* of quality and acceptability are not always clear. Many investigators collect flavor, appearance, and textural acceptability data as well as even more detailed information from consumer judges but then complain that consumers are not reliable panelists. Some suggest that consumer panelists can evaluate a wide variety of characteristics (Moskowitz, 1994) while others recommend that they be confined to judgments on acceptability alone. Perhaps the most cogent comment on this topic is that, "Consumers mostly know and can say what they like and dislike, even when they find it difficult to say why" (Conner, 1994, p. 170). The most promising alternative is to have consumers evaluate products only for overall acceptability (Conner, 1994) while simultaneously using a descriptive panel to quantify specific descriptive notes (Shewfelt et al., 1997).

Assessment tools to differentiate between defect and excellence orientations of quality characteristics are needed when it is recognized that not all characteristics are perceived equally. Some definitions of quality are clearly defect-based with the emphasis of quality management on elimination of defects (Crosby, 1979; Juran and Gryna, 1980), while others permit an appreciation of degree of excellence (Kramer and Twigg, 1970; Deming, 1986; Surak and McAnelly, 1992; Bounds et al., 1994). With respect to fruits and vegetables, appearance and textural characteristics tend to be related to a presence or absence of defects, while flavor characteristics can be expressed both in the form of defects (off-flavors) and degree of excellence (intensity of full or ripe flavor) (Shewfelt, 1993). While Hedonic (Peryam and Pilgrim, 1957) or other types of scaling are appropriate for an excellence-based quality orientation, they are probably not appropriate for defect-based quality decisions where quality is either acceptable or unacceptable. Use of a 3-point acceptability scale (unacceptable-acceptable-superior) (Shewfelt et al., 1997) or 5-point purchase acceptability scale (Moskowitz, 1994) provides a means of integrating across defect and excellence, particularly if the data are transformed and expressed as probability distributions as described above.

Factors intrinsic to the product must be *separated from those intrinsic to consumers.* One of the severest criticisms of the use of consumer data is that the consumer is not a monolith but a confederation of many different tastes and preferences. Current sensory practice tends to provide answers for the average consumer without providing an understanding of the population. Segmentation of the population is a way to overcome this problem (Land, 1988). Segmentation by demographic variables has been rejected by Moskowitz (1994), who prefers to segment by quality preferences. For this approach to work, it must be assumed that there is a finite (preferably four or less) number of distinct quality-preference segments for a given product. In addition, it should be recognized that some "factors that drive acceptability" (p. 107), are in the realm of sensory characteristics intrinsic to the food, while others are "image characteristics" (p. 447), which are related to marketing and other intangibles that are beyond the capability of most sensory laboratories (Moskowitz, 1994, p. 447). Segmentation of the consuming population can be achieved by evaluating for overall acceptability using a 3-point scale. Acceptability can be expressed as a probability distribution for each segment (Shewfelt et al., 1997).

In summary, consumer-based quality evaluation is possible. Such studies require the use of a common set of terms, must be designed to be both externally and internally valid, and must differentiate between characteristics that are based on defects and those that are based on excellence. Confining consumers to narrow choices on acceptability alone and expressing the data in the form of a probability distribution within a market segment provides an assessment tool of acceptability that is quantitative. When combined with sensory descriptive analysis, this approach (known as quality enhancement) can relate product-based quality characteristics to consumer acceptability.

QUALITY ENHANCEMENT—THE MODEL

Quality enhancement (QE) is a theoretical model that has been described previously (Shewfelt, 1994; Shewfelt and Phillips, 1996; Shewfelt et al., 1997; Malundo et al., 1997), but there have been only a few application studies. The model was developed on the premise that it is the consumer, not the researcher or the commercial distributor, who defines quality. It seeks to address concerns described in the previous section, particularly with respect to achieving both internal and external validity. The theoretical basis of the model is presented in this section of the chapter, which will be followed by examples of its application. Market segmentation by quality preference and use of a probability distribution helps overcome the difficulties encountered by variation between consumers. The QE model provides a framework for identifying critical quality attributes; generates a measurement of success with a consumer acceptability equation based on a probability distribution for each quality-preference segment; and forges linkages between basic, applied and developmental research. Quality becomes the primary focus of the research effort rather than merely an evaluative tool for some other activity. Some critical assumptions are made in applying the quality enhancement model: (1) quality characteristics are intrinsic to the fruit or vegetable, (2) acceptability involves consumer reaction to a fruit or vegetable with a given set of characteristics, and (3) contribution of a quality characteristic to consumer acceptability is a more important selection criterion than accuracy and precision of the measurement technique.

Adopting a QE perspective provides some profound consequences for viewing quality within the postharvest chain: (1) meaningful evaluation of acceptability is performed only at purchase and consumption by the consumer, thus (2) quality and acceptability cease to become a continuous function over time and become discrete functions at purchase and consumption, suggesting that (3) management decisions made by the distributor as well as by the consumer are major factors in enhancing food quality. The most striking consequence of all is that enhancement of quality rather than mere maintenance of quality is possible because the consumer is not comparing quality and acceptability of the product in hand with that at harvest. Rather, the consumer is comparing quality and acceptability with an internal level of expectations. Postharvest or distribution management techniques that can improve the quality as purchased or consumed by the consumer result in true quality enhancement.

Some serious limitations in the QE model include: (1) lack of consideration of price as a factor in acceptability and (2) lack of comparison with other items available. Both price (Jordan et al., 1985a, b, 1986; How, 1991; Malundo, 1996) and comparative availability (Sloof et al., 1996) affect consumer definition of quality. Also, both price and quality availability can lead to heightened or lowered expectations and willingness to purchase. Current sensory and consumer testing practice tends to overemphasize item-to-item comparison (e.g., few consumers conduct direct challenge studies between competing products). Our work in the area evaluates acceptability at a reasonable price ("reasonable" as determined by the consumer judge) and discourages comparison between samples. It tends to underestimate the mental comparisons that consumers are likely to make based on price and available product. At present it would appear that it is more important to clearly establish a means of assessing acceptability than to merely focus on improving accuracy and precision of quality characteristics that may have no relevance to consumer acceptance. If, however, standards can be developed in the absence of price and cross-comparisons, then the same approach could be used to determine value (quality and price) in the context of current availability.

The most pressing research needs are as follows: (1) development of consumer-based quality and acceptability studies with particular reference to determining the quality characteristics that drive acceptability; (2) comprehensive tests of the quality enhancement model to assess the usefulness (verify, refute or suggest modifications in the approach); and (3) in the longer term expansion of the model to incorporate price and availability.

APPLICATION OF THE QUALITY ENHANCEMENT MODEL

There has been no comprehensive test of the quality enhancement model, but components have been reported on mangoes, peaches (Malundo, 1996), tomatoes (Baldwin et al., 1995; Tandon, 1997), shrimp (Malundo et al., 1997) and citrus-based beverages (Reed, 1998). A brief composite of what has been learned thus far for these various commodities will be presented followed by an overview of previously unreported data collected during consumer testing of bananas.

The critical characteristics that drive acceptance of mangoes and peaches were identified using focus groups (Malundo, 1996). The most important purchase quality characteristics cited for peaches and mangoes were color, size, firmness and aroma. Flavor, mouthfeel and juiciness were important consumption quality characteristics for both fruits but flesh color and fibrousness were also important for mango consumption. Subsequent testing of preferences of consumers suggest that texture is more important to acceptability and flavor is less important than stated in focus groups.

Consumer acceptability equations were derived for tomatoes for both purchase and consumption quality characteristics using a consumer panel to judge acceptability on a 3-point scale (tastes great, acceptable, unac-

FRUIT AND VEGETABLE QUALITY

ceptable) and a descriptive panel to quantify the critical quality characteristics (Baldwin et al., 1995). Flavor acceptability equations were derived for fresh mango puree (Malundo, 1996) and citrus beverages (Reed,
1998) using similar methodology. In only the last case were separate
equations derived for different segments. A reasonably clear distinction
was achieved between purchase and consumption acceptability of six
tomato cultigens, but less success was achieved in distinguishing the flavor of mango purees. Segmentation will be necessary in flavor studies
and the quality enhancement model is not good at determining small differences between samples (Reed, 1998). Its utility will be in quantifying the role of specific characteristics in acceptability with the idea of
improving acceptability.

In studies with bananas, participating consumers were presented with
four hands representing differing maturity levels and asked to identify
the hand they would be most likely to purchase and least likely to purchase as well as the most and least likely to consume. As expected, stages
4 (green tips) and 5 (more green than yellow) were most likely to be
purchased while stage 7 (brown spots) was least likely to be purchased
(Table 8.1). Consumers clearly buy their bananas with the idea of stor-
ing them as the preferred ripeness for consumption shifts from stages
4–5 for purchase to stages 5–7 for consumption. It is an interesting di-
chotomy that 43% were least likely to consume stage 4 fruit while a
similar percentage (39%) were least likely to consume stage 7 bananas.

At another station, consumer judges were presented with three samples of cut flesh representing different maturity stages (5–8) and asked
to rate the consumption acceptability (1 = unacceptable, 2 = acceptable,

Table 8.1. Effect of Maturity Stage on Purchase Acceptability and
Willingness to Consume Fresh Bananas by 228 Participating Consumers.

Maturity	Purchase Acceptability (%)		Willingness to Consume (%)	
	Best	Least	Best	Least
4	38a	8b	8b	43a
5	42a	5b	29a	9b
6	18ab	12b	35a	9b
7	2b	75a	31a	39ab

a,b—numbers in columns followed by the same letter are not significantly different (p < 05)

3 = superior) of each sample without seeing the peel. There was a slight preference as the maturity level increased from stage 5 (green tips) to 8 (mostly brown), with the differences attributable to an increased percentage of consumers finding stage 8 fruit acceptable (Table 8.2). No significant differences were seen by brand type for consumption acceptability or percent of consumers finding the bananas superior.

In an exit survey consumers indicated that the most common method of home storage is at room temperature on the kitchen counter or table (71.9%). Other reported storage strategies included use of a ripening bowl (20.1%), in the refrigerator (6.6%) and on a banana hook (4.8%). Most participants expected a home storage life of 2–4 days (66.7%) with an additional 28.1% looking for 5–7 days of storage in the home. Only 3.5% had expectations of less than 2 days while 1.7% desired more than 7 days. Most consumers were generally satisfied with bananas available in the supermarket (80.3%) with 6.1% dissatisfied and the remaining 13.6% finding occasional dissatisfaction, particularly with the availability of fruit with the desired stage of ripeness. Slightly less of the participant population (76.3%) was generally satisfied that the bananas stayed fresh at home until they ate them, with 14.5% expressing dissatisfaction and the remaining 9.2% occasionally dissatisfied with home ripening. As expected, color (85.5%) was the primary purchase quality characteristic followed by no bruises (45.2%), firmness (28.5%), size (8.3%) and aroma (1.8%). Unexpectedly, texture (61.8%) was the predominant consumption quality characteristic followed by sweetness

Table 8.2. *Effect of Maturity Stage on Consumption Acceptability of Fresh Bananas by 228 Participating Consumers.*

Maturity	Consumption Acceptability	% Superior	% Acceptable
5	1.81 c	8	73b
6	1.93 b	16	79ab
7	1.94 b	17	76ab
8	2.01 a	17	84a
		ns	

Consumption acceptability (1 = unacceptable, 2 = acceptable, 3 = superior)
% acceptable includes those classified as superior
a,b,c—numbers in columns followed by the same letter are not significantly different ($p < 05$)
ns—not significant ($p < 05$)

(30.7%), "tastes like a banana" (26.8%) and color (11.0%). Note that multiple responses were allowed for methods of home storage, purchase quality characteristics and consumption quality characteristics.

The results from the consumer acceptability tests suggest that less than 25% of purchasers of bananas are concerned about freshness extension of bananas and that 70% have shelf-life expectations of four days or less. The emphasis on texture (primarily a concern for mushiness) as a consumption characteristic is interesting but was not verified by the consumption acceptability tests, which suggested that consumers value sweetness more than they are willing to admit. The data also show that consumers prefer consuming a riper banana than they would choose to consume based on color. Those consumers who rejected the riper bananas (stages 6 and 7) during the consumption acceptability tests cited mushiness as the reason for rejection, but for most participants color clearly predominated over texture.

FUTURE DIRECTIONS

Popularity of fresh fruits and vegetables has increased in recent years because of their image as healthy, nutritious foods. With a current trend towards dietary supplements or nutraceuticals and prospects for trends not yet contemplated, a healthy image might not be enough to improve or even maintain current consumption levels. Marketers and distributors of fresh fruits and vegetables will need to anticipate consumer desires particularly with regard to sensory quality and convenience. Current tools available to postharvest physiologists and technologists do not provide the ability to identify consumer desires or to develop new ways to help meet them. The complexity of understanding the consumer has discouraged attempts at even approaching the problem. This chapter has subdivided the problem into six distinct aspects and has argued that each aspect is soluble by focused attention. Further, it offers the QE model as a means to addressing this issue and has provided some examples of its use and application. The model as originally conceived deliberately ignores economic issues, which are critical to evaluating viability of new techniques. Although limited testing has revealed some weaknesses in detecting small differences between samples, the QE model does offer a potential bridge from the current product-oriented quality that lacks external validity to an economic-based system that will be both internally and externally valid.

REFERENCES

Baldwin, E. A., Scott, J. W., and Shewfelt, R. L. 1995. Quality of ripened mutant and transgenic tomato cultigens. *Tomato Quality Workshop Proc.*, Davis, CA.

Bounds, G., Yorks, L., Adams, M., and Ranne, G. 1994. *Beyond Total Quality Management: Toward the Emerging Paradigm*, McGraw Hill, New York.

Clydesdale, F. M., Ho, C-T., Lee, C. Y., Mondy, N. I., and Shewfelt, R. L. 1991. The effects of postharvest treatment and chemical interactions on the bioavailability of ascorbic acid, thiamin, vitamin A, carotenoids, and minerals. *Crit. Rev. Food Sci. Nutr.*, 30: 599–638.

Conner, M. T. 1994. An individualised psychological approach to measuring influences on consumer preferences, in *Measurement of Food Preferences*, H. J. H. MacFie and D. M. H. Thompson, eds., Blackie Academic and Professional, London, pp. 167–201.

Crosby, P. B. 1979. *Quality Is Free*, Mentor Books, New York.

Deming, W. E. 1986. *Out of the Crisis*, MIT Center for Advanced Engineering Study, Cambridge, MA.

Fennema, O. R. (Ed.) 1996. *Food Chemistry*, 3rd Edition, Marcel Dekker, Inc., New York.

Hendrickx, M., Maesmans, F., De Cordt, S., Nornha, Van Loey, A., Willcox, F., and Tobback, P. 1994. Advances in process modeling and assessment: The physical mathematical approach and product history integrators, in *Minimal Processing of Foods and Process Optimization—An Interface*, R. P. Singh and F. A. R. Oliveira, eds., CRC Press, Boca Raton, FL, pp. 315–336.

How, R. B. 1991. *Marketing Fresh Fruits and Vegetables*, AVI/Chapman and Hall, London.

Jordan, J. L., Shewfelt, R. L., Prussia, S. E., and Hurst, W. C. 1985a. Pricing quality attributes at the wholesale level. *J. Food Dist. Res.*, 16(2): 11–15.

Jordan, J. L., Shewfelt, R. L., Prussia, S. E., and Hurst, W. C. 1985b. Estimating implicit marginal prices of quality characteristics of tomatoes. *Southern J. Agric. Econ.*, 17(2): 139–146.

Jordan, J. L., Shewfelt, R. L., Thai, C. N., and Prussia, S. E. 1986. Transporting perishable commodities: The economic impacts of separating ethylene generators from ethylene-sensitive produce. *Southern J. Agric. Econ.*, 17(2): 139–146.

Juran, J. M. and Gryna, F. M. 1980. *Quality Planning and Analysis*, 2nd. ed., McGraw Hill, New York.

Kays, S. J. 1991. *Postharvest Physiology of Perishable Plant Products*, van Nostrand Reinhold, New York.

Kramer, A. and Twigg, B. A. 1970. *Quality Control for the Food Industry*, 3rd edition. Vol. 1, AVI/ Chapman and Hall, London.

Land, D. G. 1988. Negative influences on acceptability and their control, In *Food Acceptability*, D. M. H. Thompson, ed., Elsevier, New York, pp. 475–483.

Lawless, H. T. and Heymann, H. 1998. *Sensory Evaluation of Food: Principles and Practices,* Chapman and Hall, London.

Lyon, G. B., Robertson, J. A. and Meredith, F. I. 1993. Sensory descriptive analysis of cv. Cresthaven peaches—Maturity, ripening and storage effects. *J. Food Sci.,* 58: 177–181.

Malundo, T. M. M. 1996. *Application of the Quality Enhancement (QE) Approach to Mango* (Mangiferainda *L.*) *Flavor Research,* University of Georgia, 134 pp.

Malundo, T. M. M., Shewfelt, R. L., McWatters, K. H., and Hung, Y-C. 1997. Quality enhancement, in *Quality in Frozen Food,* M. C. Erickson and Y-C. Hung, eds., Chapman and Hall, New York, pp. 460–477.

Moskowitz, H. R. 1994. *Food Concepts and Products: Just-in-Time Development,* Food and Nutrition Press, Trumbull, CT.

Pendalwar, D. S. 1989. Modeling the Effect of Ethylene and Temperature on Physiological Responses of Tomatoes Stored Under Controlled Atmosphere Conditions, M.S. Thesis, University of Georgia.

Peryam, D. R. and Pilgrim, F. J. 1957. Hedonic scale method of measuring food preferences. *Food Technol.,* 11(9): 568–570.

Pirsig, R. M. 1974. *Zen and the Art of Motorcycle Maintenance,* Bantam Books, New York.

Reed, C. A. 1998. Sensory Techniques to Enhance Flavor Acceptability of a Citrus Flavored Beverage Using the Quality Enhancement (QE) Model, Masters of Science Thesis, University of Georgia.

Shewfelt, R. L. 1986. Postharvest treatment for extending shelf-life of fruits and vegetables. *Food Technol.,* 40(5): 70–89.

Shewfelt, R. L. 1993. Measuring quality and maturity, in *Postharvest Handling: A Systems Approach,* R. L. Shewfelt and S. E. Prussia, eds., Academic Press, Orlando, FL, pp. 99–124.

Shewfelt, R. L. 1994. Quality characteristics of fruits and vegetables, in *Minimal Processing of Foods and Process Optimization—An Interface,* R. P. Singh and F. A. R. Oliviera, eds., CRC Press, Inc., Boca Raton, FL, pp. 171–189.

Shewfelt, R. L. 1999. What is quality? *Postharvest Biol. and Technol.,* 15: 197–200.

Shewfelt, R. L., Erickson, M. E., Hung, Y-C., and Malundo, T. M. M. 1997. Applying quality concepts in frozen food development. *Food Technol.,* 51(2): 56–59.

Shewfelt, R. L. and Phillips, R. D. 1996. Seven principles for better quality of refrigerated fruits and vegetables, in *New Developments in Refrigeration for Food Safety and Quality,* W. E. Murphy and M. M. Barth, eds., ASAE, St. Joseph's, MI, pp. 231–236.

Sloof, M., Tijskens, L. M. M., and Wilkinson, E. C. 1996. Concepts for modeling the quality of perishable products. *Trends Food Sci. Technol.,* 7: 165–171.

Surak, J. G. and McAnelly, J. K. 1992. Educational programs in quality for the food processing industry. *Food Technol.,* 46(6): 88–95.

Tandon, K. S. 1997. Odor Thresholds and Flavor Quality of Fresh Tomatoes (*Lycopersicon esculentum* Mill.), Masters of Science thesis, University of Georgia.

Taoukis, P. and Labuza, T. P. 1996. Summary: Integrative concepts, in *Food Chemistry*, O. R. Fennema, ed., 3rd Edition, Marcel Dekker, Inc., New York, pp. 1013–1042.

Tucker, G. A. and Grierson, D. 1987. Fruit ripening, in Vol. 12 of *Biochemistry of Plants: A Comprehensive Treatise*, P. K. Stumpf and E. E. Conn, eds., Academic Press, New York, pp. 265–318.

van Trijp, H. C. M. and Schifferstein, H. N. J. 1995. Sensory analysis in marketing practice: Comparison and integration. *J. Sens. Studies*, 10: 127–147.

Walters, C. G. and Bergiel, B. J. 1989. *Consumer Behavior: A Decision Making Approach*, South-Western Publishing, Cincinnati, OH.

CHAPTER 9

Consumer Preference

ZAINUL ANDANI
HAL J. H. MACFIE

INTRODUCTION

THE interaction between consumers and food is a matter of interest to governments because of the importance of agriculture to rural economies, and food habits in determining health and life expectancy of a nation. It is of interest to industry because of the size of the food chain in relation to the national economy.

If we wish to understand the relationships consumers have with food, we need to enter into the world of consumer science. Consumer science in relation to food draws from a number of disciplines, primarily: sensory science, psychology, nutrition with support from statistics, physics, chemistry and biochemistry. The primary area is sensory science, which is a scientific discipline used to evoke, measure, analyze and interpret the reactions perceived by the senses of sight, smell, taste, touch and hearing to those characteristics of materials such as food (Stone and Sidel, 1993). Identifying the product qualities that are important to the consumer can be achieved by using sensory analysis (Stone et al., 1991).

In this chapter we will review some of the major influences that impact on food choice. We will examine attitudes and beliefs held by the individual and the effect of mood and post-digestion. We will then turn our attention to specific issues concerning fruit and vegetable quality using a number of sensory techniques: chewing behavior, preference mapping, the repertory grid method and expectations analysis. Finally,

158

we will look at individual differences using models that explain segmentation.

DETERMINANTS OF FOOD CHOICE

Attitudes and Beliefs

We need to understand the process by which food choices are made.
By understanding the reasons for people's choice of foods we can attempt to change these choices and so influence dietary patterns in line
with recommendations from those involved with promoting health. Food
choice is influenced by many interrelating factors and by its very nature
is a complex issue. A number of models have been proposed to delin-
eate the effects of the likely influences (Yudkin, 1956; Pilgrim, 1957;
Khan, 1981; Randall and Sanjur, 1981; Shepherd, 1985, all in Shepherd
and Sparks, 1994). Such models contain a large number of variables that
can be divided up into those related to the food, to the individual making the choice and to the external environment; but "these models are
not quantitative. They do not attempt to explain the likely mechanisms
of action of the different factors, nor to quantify their relative importance or how they interact" (Shepherd and Sparks, 1994, pp. 202, 204).
Studying the beliefs and attitudes held by an individual and the relationship between attitudes and beliefs to the choices the individual makes
offers increased knowledge about the roles played by a number of different types of factors in food choice. One structured attitude model was
proposed by Fishbein and Ajzen (1975, in Shepherd and Sparks, 1994),
the theory of reasoned action. This theory posits that intention to un-
dertake a behavior is a key antecedent of that behavior and proposes two
components as primary determinants of intention: the attitude of the subject towards a behavior, which is itself a function of the beliefs about
the outcomes of undertaking the behavior, and the importance the subject attaches to those outcomes. The latter is termed the subjective norm.

ATTITUDE TOWARD NOVEL FOODS

We choose or reject food on the basis of intrinsic and extrinsic characteristics. Intrinsic factors are sensory attributes. An example of extrinsic factors might include risk perceptions associated with the product,
the process used to manufacture the product (Saba et al., 1998), or trust
in the risk regulations (Frewer et al., 1996). We may have a positive set

of attitudes and beliefs about particular foods; it may be important to our culture, it may be produced in a way that we find specifically acceptable, i.e., organically, without meat, and so on, or we believe that it is nutritionally beneficial.

The study of how attitudes and beliefs are developed and impact on food selection is particularly relevant in the context of understanding how consumers react to novel foods and novel production processes and why consumer risk perception is different from that of experts. In a recent study by Frewer et al. (1997), the public's concerns about genetic modification were examined. The paper concluded that the public's concerns should be incorporated into the debate about the strategic development of the technology and that scientists should address ethical concerns as well as issues of risk. Indeed a great deal of research has been conducted into the issue of consumer acceptance of genetic modification but more focused research is needed, especially as public attitudes change and new products come onto the market (Frewer and Shepherd, 1998).

Mood Effects and Post-Digestive Conditioning

There are also psychobiological reasons. Our bodies require sustenance, and our brain learns quickly what foods or beverages satisfy its desire for nutritional intake or pleasure. Recent research is showing how these post-digestive effects can be very influential in determining our preferences for particular foods and flavors, for example, caffeine, alcohol and perhaps carbohydrates (Rogers and Richardson, 1993; Rogers, 1995). Infants and young children tend to reject the flavor of black coffee, but later on in life many people develop a strong preference for this drink. Mood changes occur when alcohol or caffeine are consumed; thus it is reasonable to assume that preferences for alcoholic drinks and caffeine-containing drinks may be acquired partly due to post-digestive conditioning mechanisms.

Lloyd et al. (1994) examined the acute effects of manipulating the fat:carbohydrate ratio of individual meals. Subjects' consumed isoenergetic meals containing varying amounts of fat and carbohydrate and their subsequent mood and cognitive performance were measured. The authors found that the subjects' reaction times and mood were better in the afternoon following a medium fat-medium carbohydrate lunch vs. either a high- or low-fat lunch.

In a second study Lloyd and Rogers (1994) concluded that morning mood was improved following a low-fat, high-carbohydrate breakfast vs. either a medium- or high-fat breakfast. In both studies "optimal mood and performance were associated with the meal which was closest in macronutrient composition to the subjects' typical intake at that particular time of day." Many of these effects occurred within 30 min of eating. Rogers (1995) suggests that preabsorptive and/or early postabsorptive mechanisms are implicated. Currently, the mechanisms at work here remain unclear, although the hormone cholecystokinin (Stacher et al., 1979, in Rogers, 1995) is thought to provide part of the mechanism mediating effects of meal composition.

SENSORY ATTRIBUTES THAT INFLUENCE PREFERENCE

Preference Mapping

Preference mapping offers the potential to link consumer preference patterns to sensory properties perceived by a trained panel. Preference mapping is a form of multidimensional scaling in which all consumer preferences are represented (Carroll, 1972). In a normal preference mapping trial, 30 or more consumers are recruited to assess a number of products for overall preference on a 9-point Hedonic scale. Unlike univariate analysis, preference mapping is a multidimensional algorithm, and scores for products are not averaged over consumers, but each individual is represented on the map (Earthy, 1996). A good preference map that represents a high proportion of variance among consumers may be used to reconstruct each subject's approximate order of preference for the set of products. This is accomplished by orthogonal projection of the products onto the direction of each consumer vector (Greenhoff and MacFie, 1994).

Two types of preference mapping can be used: internal or external (Carroll, 1972). Internal preference mapping is based solely on the preference scores and provides a multidimensional representation of the stimuli. With external preference mapping the preference scores derived from the consumer panel are projected onto a map that has been derived from other quantitative information about the products. This information can be in the form of sensory descriptors, physical or chemical data (Greenhoff and MacFie, 1994). For both types of mapping the position of each consumer on the map represents their own direction of increasing preference; this is known as the Vector mode. This method has several advantages over traditional methods of analyzing preference data, not only in that no information is lost by not averaging over scores, but also in that natural segmentation of consumers over the map is illustrated (McEwan, 1988/89).

Southern Hemisphere Apples

In a study by Dalliant-Spinnler et al. (1996), the technique was applied to the consumers' perception of apple varieties from the Southern hemisphere. First a trained sensory panel derived a comprehensive descriptive sensory profile of the apples. The panel elicited a total of 62 descriptors, which were divided into the following categories: external appearance, external odor, internal odor, first bite (take one bite from the center of the hemisphere with front teeth), texture during chewing, flavor during chewing and afterswallow (15 s after swallowing). In Figure 9.1 the varieties Granny Smith, Golden Delicious and Braeburn are separated from Top Red, Fuji, Royal Gala and G/S 330 along the first principal component (PC), while Celeste and Compact Golden Delicious are well separated from Braeburn and Aurora along the second PC. In Figure 9.2 which represents the plot of the attributes, descriptors such as grassy, green apple and damp twigs occupied the same space. These

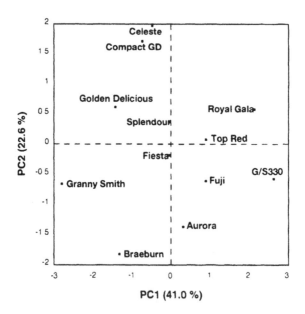

FIGURE 9.1 PCA: Product map for the first two principal components of apples (62 descriptors).

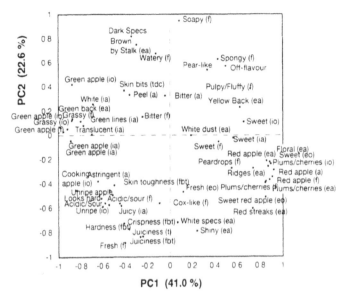

FIGURE 9.2 PCA: Descriptor map for the first two principal components of apples. The codes correspond to the descriptors used to describe the apples—(ea) external appearance, (eo) external odor, (o) internal odor, (ia) internal appearance, (fdc) flavor during chewing, (a) afterswallow.

characteristics are associated with Granny Smith; this variety is also associated with cooking apple, astringent drying and unripe apple. Characteristics typical of Braeburn are found along the second axis: juiciness, crispness, hardness and fresh. The red apples were associated with the descriptors: spongy, off-flavor, pear-like and pulpy/fluffy. Top Red and Royal Gala were also characterized as being sweet.

Hedonic evaluation by consumers using the same varieties revealed a similar picture to that of the trained panel (refer to Figures 9.3 and 9.4). Internal preference mapping revealed four dimensions of preference but we will only highlight the first two here. The 95% confidence ellipses indicated in Figure 3 show three main groupings of varieties along the first PC. The second PC separates Fiesta and Fuji and Compact Golden Delicious from their respective groupings. A plot of the sensory descriptors can be seen in Figure 9.4. The first dimension (PC) is positively correlated with hardness, crispness, juiciness, shiny appearance and fresh flavor, and is negatively correlated with soapy and spongy. The second dimension is that of sweetness. Thus Braeburn, Aurora and Granny Smith score highly on the first dimension, and to some

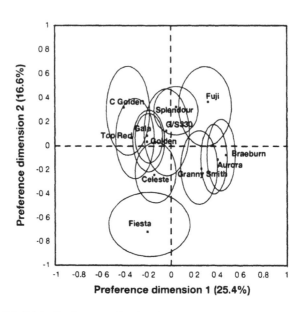

FIGURE 9.3 PCA: Product map for the first two principal components of apples.

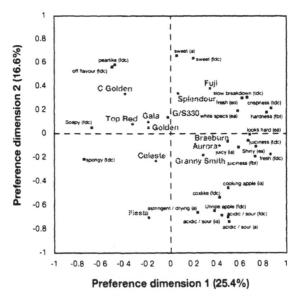

FIGURE 9.4 PCA: Descriptors significantly correlated with the first two preference dimensions.

164

extent Fuji does as well. Fiesta is not thought to be sweet but is considered unripe and acid/sour along with Braeburn, Aurora and Granny Smith. The study indicated a reliable and reproducible pattern of consumer segmentation towards either a preference for sweet, coxlike style or a hard acidic style. The market opportunity for an English variety similar to the Granny Smith is perceived.

By identifying key components of consumer preference in this way, the components can be related back to physical and chemical studies, allowing for the development of clear specifications that can be applied in fruit selection. Another method of selection or breeding may be achieved by using genetic markers. Genetic markers for mildew resistance, scab resistance and fruit color have been located and by extending this work markers may be located for texture and flavor characteristics. In the future is hoped that specific components of organoleptic quality will be related to these markers enabling the desired attributes to be selected (Alston et al., 1996) for a wide range of fruit and vegetables.

Chewing Behavior

We still do not understand enough about the mastication process, that is, how food is broken down during mastication, and how this dynamic process influences our perception of the texture and flavor of foods. In recent years electromyographs (EMG) of the masticatory muscles have taken us a step further toward answering these questions. EMG has been used to monitor the activity of the masticatory muscles during chewing using small electrodes placed on the skin over the muscles. The activity of these muscles reflects the textural quality of the food being eaten and how this changes during the mastication process. Consumers interact with food (in the mouth) in different ways, and derive different sensory information about the food they are consuming (Brown, 1995). Examination of the mastication patterns for different individuals and foods should assist in understanding the way different foods are perceived.

The authors believe too much emphasis has been put on the supposed properties of the food and not enough on the mechanisms of perception. Examining all the factors that bear on a consumer in choosing and subsequent consuming foods and beverages, it is clear that previous research on fruit and vegetable quality has not acknowledged these adequately. Future research needs to take a far more integrated perspective of both the people factors and the food factors and the interactions between the two. For the rest of this chapter we will illustrate some of the tools and concepts we have been using to link consumers and foods.

REPERTORY GRID—A TOOL TO ELICIT CONSUMER DESCRIPTORS AND PERCEPTIONS

The Repertory Grid Method (RGM) is another technique that has been recently used to look at apples (Bhanji et al., 1997). The method was originally developed by Kelly (1955) to identify the constructs that people use to structure their perceptions of the social world. The basis for this method is the elicitation of "constructs." A construct can be defined as a way in which two things are alike and, in the same way, different from a third (Kelly, 1955). Applying the idea of the RGM to food acceptability was first proposed by Olsen (1981).

Objects are arranged into groups of three (triads) such that each object appears in at least one triad and that one object from each triad is carried over to the next triad. The subject is asked to state in which way two objects are alike and different from a third, and descriptions for likeness and difference are recorded (Sampson, 1972). Once the subject has exhausted all his/her constructs for a given combination, the two remaining combinations within the triad are similarly presented and constructs elicited in the same way. The constructs may then be associated with 100 mm visual line scales or a box scale anchored with the extremes of the construct as described by the subject (Raats, 1992). The method has been applied in a number of food-related studies (Thomson and McEwan, 1988; and Scriven et al., 1989).

The data from the RGM can be analyzed by Generalized Procrustes Analysis. Generalized Procrustes Analysis (GPA) is an empirical multivariate method that can be used to explore the relationship between samples and attributes on a multivariate space (McEwan, 1989). When applying GPA to data, each configuration is a matrix of scores from one individual who has assessed a number of samples for a given set of attributes or constructs. Samples must be identical for each of the subjects. GPA allows each individual to have a unique set of attributes or constructs (Raats, 1992).

The data transformation involves three steps, which ensure that the results can be meaningfully compared across subjects: translation, rotation/reflection and scaling. Thus, GPA takes into account three sources of variation inherent in sensory data. Translation can remove the effect of individual panelists who constantly under- or overscore a particular attribute. Another source of variation occurs when subjects interpret the same term differently. Rotation/reflection increases the similarity between the configurations (Piggott, 1986). Finally, isotropic scaling is used to overcome the problem of subjects varying in their range of scoring. One subject may use a very small range of the scale, while another may use a much larger range to express differences between samples on a particular attribute. GPA uniformly rescales individual assessor configurations to a common overall value to eliminate this source of variation (McEwan, 1989).

MEALINESS—A MULTILINGUAL VOCABULARY

The consumption, nutritional contribution and sensory attributes of fruit and vegetables make them important foodstuffs. Variation in their texture has a great deal to do with their acceptability. Mealiness, in general, is regarded as a negative characteristic except in cases where it is a calculated aim. On the whole, it appears to be of low interest and appeal (Szczesniak et al., 1971).

In the study by Andani et al. (1998), three commercially important varieties of "fresh" dessert apples: Jonagold, Cox and Schone van Boskoop, known to be susceptible to varying degrees of mealiness, were evaluated. The aim of this work was to establish how consumers define mealiness and explore crosscultural differences among European countries. The perception of mealiness among 25 consumers in five consumer groups, the U.K., Spain, Belgium (Flemish, French) and Denmark, was explored by the RGM and the data were analyzed by GPA.

Figure 9.5 shows the consensus plot from the repertory grid study. The consumers were all subjected to the same GPA in order to look at differences in perception across consumers in the different countries. To explain the differences between perceptions, it was necessary to plot the sample mean for the consumers within each country and then look at their distribution around the global mean, which portrays how different they are. From the map it can be seen that no one consumer group differed significantly from the others in terms of how they perceived the individual samples. The position of the groups around each sample is fairly tight. Second, consumers perceived the differences between sam-

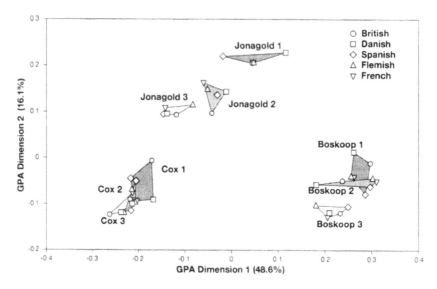

FIGURE 9.5 GPA consensus plot from repertory grid study of apple mealiness. Showing variations among the different consumer panels: 1 = fresh, 2 = midpoint, 3 = mealy.

ples quite similarly. Overall there is no structure in the way that the consumer panels are positioned around each sample. It can therefore be concluded from this plot that the consumers perceived the differences between the samples similarly. This led us to conclude that there is a cross-cultural consensus with respect to the perception of mealiness. Having established this cross-cultural perception it was necessary to look at how the consumers described their perceptions of the apples.

The consumers did vary in how they used the descriptors. Interest lay in exploring these differences in more detail. Were there differences between consumers within one panel in how they used these descriptors and how they described the differences? In order to help answer this question a curved box plot procedure was developed. Figure 9.6 shows the curved box plots for the British and Danish consumers.

The curved box plot is similar to a normal box plot but here it is the angular distribution of the consumer descriptors that is represented. The box itself accounts for 50% of the attributes and the line inside signifies the direction of the median attribute vector. Lines or whiskers extend up to 1.5 times the interquartile range, or to the left-most or right-most attribute vector. Attributes outside of the interquartile range are represented by an asterisk and can be considered outlying. The use of the descriptors "crisp" and "mealy" by Danish and British consumers is

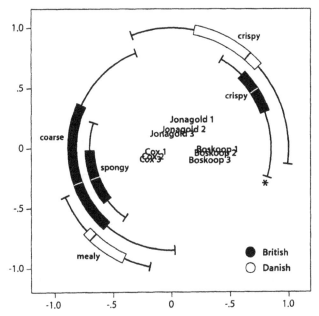

FIGURE 9.6 Curved box plots of the crisp and mealy descriptors used by British and Danish consumers.

shown in Figure 6. Most of the Danish consumers found Jonagold apples to be crisp, although a few associated it with Boskoop. A similar variation was observed among the British consumers. The Danish consumers described the texture of Cox apples as mealy, while the British described it as coarse and spongy.

GPA suggested a crosscultural consensus with respect to consumer perception of mealiness in apples. The descriptor mealy was well understood by Danish consumers, but not British consumers, who used a wider variety of descriptors like coarse and spongy. Future work will establish similar vocabularies for consumers in Belgium and Spain and establish a crosscultural vocabulary for consumer perception of mealiness.

Although there appears to be an increasing globalization of consumer patterns within the food area, Askegaard and Madsen (1995) believed that major national and regional differences persist. This did not appear to be so in this study, however; but it is important to highlight here that the use of the crosscultural approach in the case of food has considerable cultural significance (Askegaard and Madsen, 1995; Fischler, 1990; Mennell et al., 1992), thus emphasizing that the results from one country may not be readily generalized to other countries (Grunert, 1997).

EXPECTATION ANALYSIS—A TOOL TO IMPROVE QUALITY PERCEPTION AND INFLUENCE CHOICE

Defining consumer perception of quality as matching expectations shows us how giving improved information about products can increase perception of quality faster and more cheaply than breeding new varieties. Before illustrating this idea with a couple of examples, let us first consider the concept of expectations in relation to fresh fruit and vegetables.

Consumers have expectations about the performance of a product. In the case of fruit and vegetables as with any other food products, it plays a very important role as it may improve or degrade the perception of the fruit or vegetable product before it is even tasted (Deliza, 1996). The consumers' expectations are supported by information and/or by previous experience. In the case of fresh fruit and vegetables, consumers base their expectations to a large extent on external attributes. We will now illustrate the concept of expectations with a study by Deliza (1996) using passion fruit juice.

A Model to Demonstrate the Role of Expectations

Deliza (1996) constructed a hypothetical model of the role of expectations (refer to Figure 9.7). A flow diagram has been used to demonstrate the role of expectations in product selection and evaluation and repeated purchase. As noted earlier, consumers have previous experience or information, which lead to them having prior expectations. Expectations are then created based on these prior expectations and those created by the product itself, along with any label, package, ads, price, and so on (and color, size, shape, etc., in the case of fresh fruit and vegetables). These expectations can be low, which leads to product rejection, or high, which is likely to lead to the product being chosen. The chosen product is then tasted and the expected sensory attributes of the product are either confirmed or disconfirmed. Confirmation leads to satisfaction and most probably repeated product use. Disconfirmation can also lead to satisfaction—known as "positive disconfirmation." A negative confirmation of expectations can also occur, and this will probably lead to the product being rejected. Finally, the model predicts that confirmation or disconfirmation will affect the next experience with the product, which will then contribute to either raise or lower the con-

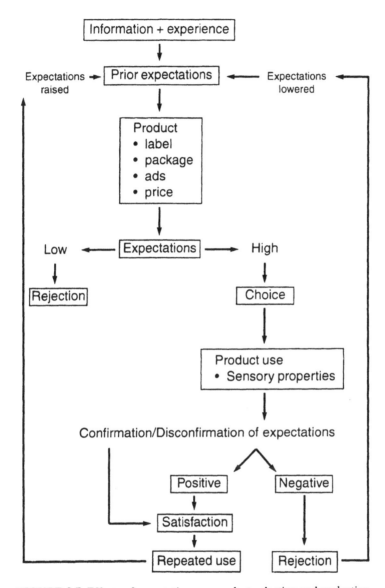

FIGURE 9.7 Effects of expectations on product selection and evaluation.

sumer's expectations. So it follows that if a consumer is well informed about the internal sensory characteristics of a product, then the consumer's expectations will be satisfied and this will lead to acceptance of the product (Escamilla-Santana, 1994).

Packaging—Influence on Expectations

Deliza (1996) also explored the way in which packaging may influence the consumer's expectations with an unfamiliar product. Following preliminary trials to establish which were the salient attributes of the unfamiliar product (consumers who had never tasted passion fruit juice were recruited for this study), a large-scale consumer trial was conducted. Consumers were first asked to evaluate the expected attributes (sweetness, sourness and refreshingness) and liking. They were then given passion fruit juices and asked to rate the actual attributes having been told that the actual attributes they were rating were from the packaging they had been shown. The results indicated that the features of the packaging contributed to establishing the expectation relating to the sensory attributes and liking for passion fruit juice. The manipulations appeared to increase the expected "freshness" of the product, most notable of these were the use of a white background color as opposed to a orange background color and lots of information on the packaging; using a major brand name and representing the passion fruit with a drawing as opposed to a photograph had a smaller effect. Packaging that contained lots of information increased the expected "naturalness" of the product while no information had the opposite effect. The effect of information on the packaging also had a marked effect on the expected "pureness" of the product. Again, lots of information on the package increased the effect of expected "pureness" while no information decreased the effect.

INDIVIDUAL DIFFERENCES ANALYSIS— MODELS THAT EXPLAIN SEGMENTATION

Another factor that we should take into account when considering research using consumers is that of individual differences. This area of research is too large to bring into this chapter and merits a chapter of its own. Instead we wish to highlight three measurement techniques: the Elaboration Likelihood Model, the Private Body Consciousness test and the Need for Cognition test.

The Elaboration Likelihood Model

The Elaboration Likelihood Model (ELM) is used to explain attitude change after information has been processed (Petty and Cacioppo, 1986).

According to this model, there are two routes to processing cognitive persuasion: the central route and the peripheral route. A consumer following the central cognitive processing route evaluates a product using relevant information, while a consumer following the peripheral route gives more weight to simple inferences or cues in his or her final judgment than the actual product attributes.

The Private Body Consciousness Test

The Private Body Consciousness (PBC) test assesses how aware subjects are of their internal sensations (Miller et al., 1981) by asking questions on how sensitive they are to changes in body temperature, internal tensions, heart rate, dryness of mouth and throat, and hunger contractions. The individuals are classified into having high or low PBC based on the answers given to these questions.

In a recent study by Jaeger et al. (1998), sensory preference was investigated among British and Danish consumers using apples that had been specifically selected for their susceptibility to varying degrees of mealiness. High-PBC individuals appeared to rely more strongly on the sensory characteristics of the apples when forming their preference ratings than low-PBC individuals.

The Need for Cognition Test

Need for Cognition (NFC) has also been used to demonstrate that it moderates the route information is processed (Inmam et al., 1990). The NFC test assesses the tendency for an individual to engage in and enjoy thinking (Cacioppo and Petty, 1982). Questions on mood, fatigue, involvement, and time availability are asked. High-NFC individuals tend to engage in more cognitive activities and are more likely to process additional information, while low-NFC consumers are more likely to take the peripheral route. Deliza (1996) in her study on expectation and instant coffee concluded that high-NFC subjects were less affected by the "label," and therefore by expectation, than low-NFC subjects. In her passion fruit juice study she surmised that individuals with a low NFC considered packaging with a white background to have a sharper taste and packaging with an orange background to be less sharp. High-NFC individuals appeared to rely more strongly on the sensory characteristics of the juices when rating the level of "sharpness."

CONCLUSIONS

More and more emphasis is being placed on the consumer and the marketplace as companies invest more resources in this area because they are realizing the importance of the consumer is to the boyancy of their business. More and more pressure is being placed on learning more about consumer behavior, especially why a product is liked and selected (Stone et al., 1991). By determining those product attributes that are most important to consumer, we can focus development efforts on those attributes to ensure they are optimal, and thus better satisfy consumer expectations. With the consumer's growing interest in quality, it is in the best interests of industry to be more sensitive to consumer needs, particularly as it relates to product's sensory quality." We have been given the tools to provide a bridge between marketing and R&D and we must use them to minimize product failure and give better targets to breeders. By integrating the breeders, producers, retailers and "the consumer" we will have greater choice, better quality and better value for the money.

ACKNOWLEDGMENTS

The authors would like to thank their colleagues at the Institute of Food Research, especially Drs. Brown, Frewer, Rogers, Shepherd and Sparks for their input into the content of this review. Parts of the work reported here were funded by the EU FLAIR project CT95–0302.

REFERENCES

Alston, F. H., Evans, K. M., King, G. J., MacFie, H. J. H., and Beyts, P. K. 1996. The potential for improving organoleptic quality in apples through marker assisted breeding related to consumer preference studies, in *Agri-Food Quality. An Interdisciplinary Approach*, G. R. Fenwick, C. Hedley, R. L. Richards and S. Khokhar, eds., The Royal Society of Chemistry, London.

Andani, Z., MacFie, H. J. H., and Wakeling, I. 1998. Mealiness in apples: towards a multilingual vocabulary (poster 11578). 3rd Pangborn Sensory Science Symposium, Sense & Sensibility, Norway, 9–13, August.

Askegaard, S. and Madsen, T. K. 1995. European food cultures: an exploratory analysis of food related preferences and behaviour in European regions. MAPP Working Paper No. 21. The Aarhus School of Business, Aarhus, Denmark.

Bhanji, Z., Jaeger, S. R., Wakeling, I., De Smedt, V., and Gomez, C. 1997. Consumer perception of mealiness in dessert apples across a span of EU countries. *Proceedings of the Sixth Food Choice Conference*, 207.

Brown, W. E. 1995. The use of mastication analysis to examine the dynamics of oral breakdown of food contributing to perceived texture, in *Characterisation of Food: Emerging Methods*, A. G. Gaonkar, ed., Elsevier Science B.V., London, pp. 309–326.

Cacioppo, J. T. and Petty, R. E. 1982. The need for cognition. *Journal of Personality and Social Psychology*, 42(1): 116–131.

Carroll, J. D. 1972. Individual differences and multidimensional scaling, in *Multidimensional Scaling: Theory and Applications in the Behavioral Sciences*. Vol. II, R. N. Shepherd, A. K. Romney and S. B. Nerlove, eds., Seminar Press, New York.

Dalliant-Spinnler, B., MacFie, H. J. H., Beyts, P. K. and Hedderley, D. 1996. Relationships between perceived sensory properties and major preference directions of 12 varieties of apples from the southern hemisphere. *Food Qual. Pref.,* 7(2): 113–126.

Deliza, R. 1996. The Effects of Expectation on Sensory Perception and Acceptance. Ph.D. Dissertation, Univ. of Reading, U.K.

Earthy, P. 1996. Personal communication. Context Effects on Preference and Preference Mapping. Ph.D. Dissertation, Univ. of Reading, U.K.

Escamilla-Santana, C. 1994. A Method to Develop Sensory Quality Standards to Maximize Consumer Acceptance. Ph.D. Dissertation, Univ. of Reading, U.K.

Fischler, J. C. 1990. *L'Omnivore*, Odile Jacob, Paris.

Frewer, L. J., Howard, C., Hedderley, D., and Shepherd, R. 1996. What determines trust in information about food-related risks? Underlying psychological constructs. *Risk Analysis,* 16: 473–486.

Frewer, L. J., Howard, C., and Shepherd, R. Winter 1997. Public concerns in the United Kingdom about general and specific applications of genetic engineering: risk, benefit and ethics. *Science, Technology, & Human Values,* 22(1): 98–124.

Frewer, L. J. and Shepherd, R. 1998. Consumer perceptions of modern food biotechnology, in *Genetic Engineering for the Food Industry: A Strategy for Food Quality Improvement*, S. Roller, ed., Blackie Academic, New York, pp. 27–46.

Greenhoff, K. and MacFie, H. J. H. 1994. Preference mapping in practice, in *Measurement of Food Preferences*, H. J. H. MacFie and D. M. H. Thomson, eds., Blackie Academic & Professional, London, pp. 137–166.

Grunert, K. G. 1997. What's in a steak? A cross-cultural study on the quality perception of beef. *Food Qual. Pref.,* 8(3): 157–174.

Inmam, J. J., McAlister, L., and Hoyer, W. D. 1990. Promotion signal: proxy for a price cut? *Journal of Consumer Research,* 7(June): 74–81.

Jaeger, S. R., Andani, Z., and Macfie, H. J. H. 1998. Consumers' preferences for fresh and aged apples: a cross-cultural comparison. *Food Qual. Pref.*, 9(5), 355–366.

Kelly, G. A. 1955. *The Psychology of Personal Constructs*, Norton, New York.

Lloyd, H. M., Green, M. W., and Rogers, P. J. 1994. Mood and cognitive performance effects of isocaloric lunches differing in fat and carbohydrate content. *Physiology and Behaviour*, 56: 51–57.

Lloyd, H. M. and Rogers, P. J. 1994. Acute effects of breakfasts of differing fat and carbohydrate content on morning mood and cognitive performance. *Proceedings of the Nutrition Society*, 53: 239A.

McEwan J. A. 1988/9. Statistical methodology for the analysis and interpretation of sensory profile and consumer acceptability data. Campden Food and Drink Research Association: Technical Memorandum 536. Maff project 1843.

McEwan, J. A. 1989. Statistical methodology for the analysis and interpretation of sensory profile and consumer acceptability data. Campden Food and Drink Research Association, MAFF Project 1843, 46.

Mennell, S., Murcott, A., and van Otterloo, A. 1992. *The Sociology of Food, Eating, Diet and Culture*. Sage, London.

Miller, L. C., Murphy, R., and Buss, A. H. 1981. Consciousness of body: Private and public. *Journal of Personality and Social Psychology*, 41(2): 397–406.

Olsen, J. C. 1981. The importance of cognitive processes and structures for existing knowledge understanding food acceptance, in *Criteria of Food Acceptance*, J. Solms and R. L. Hall, eds., Forster, Zurich.

Petty, R. E. and Cacioppo, J. T. 1986. *Communication and Persuasion. Central and Peripheral Routes to Attitude Change*, Springer-Verlag New York Inc., New York, p. 262.

Piggott, J. R. (Ed.) 1986. *Statistical Procedures in Food Research*, Elsevier Applied Science Publishers Ltd., London.

Raats, M. M. 1992. The Role of Beliefs and Sensory Responses to Milk in Determining the Selection of Milks of Different Fat Content. Ph.D. Dissertation, Univ. of Reading, U.K.

Rogers, P. J. and Richardson, N. J. 1993. Why do we like drinks that contain caffeine? *Trends in Food Science and Technology*, 4: 108–111.

Rogers, P. J. 1995. Food, mood and appetite. *Nutrition Research Reviews*, 8: 243–269.

Saba, A., Moles, A., and Frewer, L. J. 1998. Public concerns about general and specific applications of genetic engineering: A comparative study between the UK and Italy. *Journal of Nutrition and Food Science*, 28(1): 19–30.

Sampson, P. 1972. Using the repertory grid test. *Journal of Marketing Research*, *IX:* 78–81.

Scriven, F. M., Gains, N., Green, S. R., and Thomson, D. M. H. 1989. A contextual evaluation of alcoholic beverages using the repertory grid method. *International Journal of Food Science and Technology*, 24: 173–182.

Shepherd, R. and Sparks, P. 1994. Modelling food choice, in *Measurement of Food Preferences*, H. J. H. MacFie and D. M. H. Thomson, eds., Blackie Academic and Professional, London, pp. 202–226.

Stone, H. and Sidel, J. L. 1993. *Sensory Evaluation Practices*, 2nd ed., Academic Press, Inc., San Diego, p. 11.

Szczesniak, A. S. and Kahn, E. L. 1971. Consumer awareness of and attitudes to food texture. *Journal of Texture Studies, 2: 280–295.*

Thomson, D. M. H. and McEwan, J. A. 1988. An application of the repertory grid method to investigate consumer perceptions of foods. *Appetite*, 10: 181–193.

CHAPTER 10

Instrumental Data—Consumer Acceptance

BERNHARD BRÜCKNER
HELGA AUERSWALD

INTRODUCTION

TECHNICAL progress has accelerated the improvement of methods to analyze food and this development probably will continue. At the same time attention is amplified, focusing on the consumer of the products. Increased competition and low success rates in new products seem to justify efforts to gain knowledge of and implement those food attributes that consumers will perceive and like.

To become able to predict the Hedonic reaction of consumers to perceivable attributes, objective, measurable data and consumer acceptance data are combined. Schutz (1990, 1993) summarized that the objective of this research for horticultural products is, besides the purely fundamental scientific, to

- develop guidelines for the production (modification and optimization of cultivation techniques and selection of appropriate cultivars)

- offer criteria for quality assurance and quality management

QUALITY CRITERIA

How can directly perceivable attributes, which can influence acceptance ratings, be integrated into the whole quality frame of consumers?

To answer this question we conducted a consumer survey in Berlin (Florkowski et al., 1996a,b). Several questions concerned the relative importance of quality criteria of vegetables (Table 10.1).

Among the most important criteria were freshness, low residues, good taste and firm, crisp texture. Size and uniformity of size were rated least important. Good taste and healthiness were given as the most important reasons to eat vegetables. Willingness to pay the highest additional price was indicated for improved flavor and production harmless to the environment. A recent European study on food choice (Lennernäs et al., 1997) reported quality/freshness, price, taste and healthiness as most important to European consumers, but family preferences, habit and convenience also were rated important.

In summary all the different criteria can be grouped into four major categories (Table 10.2). The first category includes appearance, color and texture. These attributes can be readily determined visually or by touch. Therefore, they can be used as criteria throughout the production and distribution system. The second category includes flavor and mouthfeel. As in the first category, the attributes are closely related to physical and chemical attributes and are perceivable by the human senses during consumption and by analytic instruments.

Attributes with health relevance (minerals, vitamins, fiber, bioactive substances) and safety (residues, nitrate, etc.) are not open to direct consumer evaluation, and therefore they cannot contribute to an attribute liking—hence they are not detectable by an acceptance test. Extrinsic product characteristics such as marketing variables, social and environ-

Table 10.1. Relative Importance of Quality Criteria of Vegetables.

1. Freshness
2. Residues below allowable level
3. Good taste even if not uniform
4. Firm, crisp
5. No damage
6. Harvested when ripe
7. Price
8. Clean
9. Organically grown
10. Size
11. Uniform size

Table 10.2. *Quality Criteria Categories.*

Categories	Place and Person of Perception			Methods to Assess		
	By Producer/ First Buyer	At Retail Purchase	At Consumption	Acceptance Test	Analytic Measurement	Labeled Information
1. Appearance, color, texture	•	•	•	•	•	•
2. Flavor, mouthfeel			•	•	•	•
3. Health relevance, safety					•	•
4. Social, environmental, marketing						•

mental aspects are very important and are integrated by the consumer to form a purchase decision, but are not able to be measured by instruments. Finally, the overlapping fields of instrumental investigation and acceptance testing are attributes, including purchase factors like color, size, shape, absence of defects, firmness to the touch and consumption quality attributes like flavor and mouthfeel.

ROLE OF ACCEPTANCE

The perception and liking of product attributes does guarantee a positive, single or multiple purchase decision. It is well known that a variety of influences and circumstances are involved when it comes to the decision to buy or not to buy even an acceptable product (Cardello and Sawyer, 1992; Grunert, 1995; Grunert et al., 1996; Oude Ophuis and van Trijp, 1995). What is the role of acceptance in this context? The use of the term "acceptance" varies considerably from a synonym of pleasantness to an ultimate decision. Most facets in the literature lie between the two definitions of acceptance by Amerine et al. (1965):

- Actual utilization (purchase, eating). May be measured by preference or liking for specific food item.
- An experience or feature of experience, characterized by a positive (approach in a pleasant) attitude.

Applying the first definition means measuring what individuals actually consume under normal conditions that surround eating behavior (Schutz, 1990, 1993). But acceptance measurement usually takes place in different situations, which can reduce the validity of results and can make relations between a measure of food acceptance and actual food consumed disappointing (Schutz, 1993). Therefore, an interdisciplinary approach seems to be necessary.

The second, more restrictive definition is widely used and applied in sensory acceptance tests (Meilgaard et al., 1991). The objective of these tests is to identify the liking or nonliking of a stimulus or product by naive consumers, rather than perceived intensities by trained or nontrained panelists. The tests should determine the "affective status" of a product without claiming knowledge of actual purchase behavior.

Sensory Acceptance

The background of sensory acceptance measurement are all the circumstances of the sensory perception and interpretation, including:

expectations and attitudes, selection process (preferences, habit), individual preparation (chewing, saliva), attention, perception (psychophysics), integration (peripheral, cognitive access), affective involvement, recognition, verbalization, quantification.

All of these steps have been discussed several places in detail (Andani and MacFie, 1999; Cardello and Sawyer, 1992; Brown et al., 1996; McBride and Anderson, 1990; Kroeze, 1990; Pangborn, 1981; Schutz, 1993; Conner, 1994). There does not seem to exist a comprehensive model integrating the theoretical background of the different disciplines involved. Although some of the above aspects are covered elsewhere in this book, I will describe some processes closely related to sensory consumer acceptance testing.

THE PERCEPTION PROCESS

To lead to any human sensation a perception process of a physical or chemical or combined stimulus must take place. To produce a sensation the stimulus has to be present for a certain length of time and to be above the absolute threshold. Identification of the specific stimulus usually requires a further increase of the stimulus level to reach the recognition threshold. Above the terminal threshold, no further increase in the intensity of the stimulus is perceived. A stimulus intensity difference above the difference threshold is necessary to produce a just noticeable difference (JND). This noticeable difference increases at higher stimulus levels (Weber's law), thus being usable as "natural" scaling units for sensory intensity.

All of these threshold values are not constant, but vary considerably from person to person and from time to time (Land, 1983; Meilgaard et al., 1991). Therefore care must be taken when threshold values are used to calculate flavor intensity. To use the absolute threshold as the yardstick of intensity is recommended only for 0.5–3-fold concentrations (Meilgaard et al., 1991). For practical levels of stimuli, there seems to exist no relation between threshold and suprathreshold sensitivity in theory or observation (Booth, 1990).

The sensory perception of a single stimulus has been described by psychophysical laws. The strength of a perceived sensation is calculated as a function of the magnitude of the stimulus. Fechner's law uses JNDs in a logarithmic, Steven's law uses a power form. The Beidler Model shows a sigmoidal structure (Meilgaard et al., 1991). All of them can be discussed before the common background of information theory (Norwich, 1991).

Determination of the intensity of the human sensation of taste in most cases originates from discrimination tests or descriptive rating (by categories), scoring (on line scales) or ratio scaling (magnitude estimation). In each of these cases it is words that have to express precisely what sensations have been perceived. These words have to be found (if they exist at all) and used by different assessors the same way (Thompson and MacFie, 1983). Another possible problem arises from the fact that if notes from complex food items are compared quantitatively, word profiles are compared. Possible additional effects in the overall impression brought by a combination of the parts are left aside (Thompson and MacFie, 1983).

In "real" foods like fruits and vegetables a complex of stimuli rather then a single stimulus is present. The effective stimulus in complex food is not clear. Correlation of instrumental data and sensory values is only meaningful when the instruments measure the effective stimulus. For instance, in texture research there has been substantial effort to identify the effective principles, or better, how they can be measured instrumentally (Szczesniak and Kahn, 1971; Kapsalis and Moskowitz, 1978; Bourne, 1983; Brown et al., 1996).

In this context Kroeze (1990) discusses the analytic or synthetic character of the human senses. In case of mixed taste stimuli he reports examples of integration of perceptions, a breakdown to a smaller number of sensing units. The result of these processes can be described by additive, weighted additive, enhancing or suppressive or dominant attribute models (McBride, 1990). In the last case one attribute above a certain intensity suppresses the perception of another attribute—thus being dominant. McBride and Anderson (1990) emphasize that a there may be no relationship between the psychophysical functions of individual stimuli and the mixture psychophysics.

Whatever mechanisms apply, it is not clear whether mixtures can be analyzed backward to yield single components (or stimuli) or are perceived as a whole percept, a pattern, as emphasized by Gestalt psychology. Kroeze (1990) gives the examples of the drawing of a square that does not present itself as a set of four lines or the pattern of tastes, smells and texture that is immediately recognized as cheese. In both cases the subjects would be able to analyze the percepts to some degree: into four lines or into several taste, smell and texture components. But analysis requires attentional effort, which in cases of a tightly organized percept, a strong impression of wholeness, can lead to an informational overload or being out of the reach of an attentional mechanism. Whatever the integration level is to which perception of attribute intensities is reduced, there is a further process of Hedonic integration to a final, unidimensional, Hedonic tone (McBride and Anderson, 1990). Moskowitz and Krieger (1993, 1995) quantified the relative importance of the liking of attributes in relation to the overall liking. Moskowitz et al. (1994) are among those who claim a synthesized concept as a whole, only as such adapted to consumer segments.

Although there may be some inaccessible pattern recognition, in principle, it seems possible to predict Hedonic optima on the basis of ingredients (Lawless, 1990). Scaling, precision and accuracy were emphasized by Lawless as upper limits to explainable consumer acceptability. As an additional requirement he states that the sensory characteristics and their interactive mechanisms are fully specified and quantified in the measurement procedure (Lawless, 1990), which has to be improved as the theoretical basis of acceptance evaluation seems not to be well established (McBride, 1990). It is important to point out the importance of ideal and rejection points (Booth, 1990). Consumer reactions can be calculated and should be related to operationalized measures of the attributes of the product that are actually operative in the mind of each customer making choices (Booth, 1990).

When we agree on a definition of acceptance reflecting an affective status of a product, we must separate all the situational and extrinsic influences from the perceptible properties of the product itself. Furthermore, we have to realize that the perceptible properties do not travel individually, but interact with others and form indivisible stimuli. In the following section an example is presented, where sensory effects only become explainable when different properties are understood as being effective as a unit. It is an example of sensory integration. Another example is given, where an attribute is dominant and directly affects an Hedonic reaction. We also find a situation, where the perceptions of several attributes are condensed into fewer dimensions of Hedonic impression. An example is given, where changes of color, texture and flavor properties resulted in integrated Hedonic reactions shared by one consumer segment and other reactions shared by another segment.

To identify the mechanisms of integration on the sensory and on the Hedonic level, it is necessary to have detailed information of the physicochemical product properties, the resulting sensations and Hedonic reactions. This allows for analyzing how (hopefully not too many) single or a blend of properties translate into important sensations and how single or blended sensations translate into Hedonic tones in consumers or certain groups of consumers. If we have collected some knowledge about the underlying mechanisms, we can go backwards from the acceptable Hedonic tones and try to identify those lower and upper levels of properties that combine to those sensations, which in turn combine to acceptable hedonic tones. There may be multiple combinations of properties to form an acceptable Hedonic tone for one consumer group and several different preference groups may exist in parallel.

EXAMPLES OF RELATIONS BETWEEN INSTRUMENTAL AND ACCEPTANCE DATA

In the case of fresh fruits and vegetables, there are limited data on independent, factorial experiments to evaluate consumer reactions. One example is the independent effect of sugar and acid concentrations in tomatoes to which sugars and acids were added (Malundo et al., 1995). Independently from each other, increased sugar and increased acid level led to a linearly increased sweet and sour descriptive value and influenced consumer scores. To study physicochemical attributes, as generated during production, and their relevance for the prediction of consumer liking, we conducted instrumental and sensory investigations on tomato fruits. In two experiments tomatoes were grown soilless, using nutrient film technique in the greenhouse. A trickle system applied the nutrient solution (Sonnefeld and Straver, 1988) to rockwool cubes placed in a trough.

Experiment 1

In the first experiment six different cultivars were grown: round type 'Gourmet,' 'Pronto,' 'Stamm,' longlife type 'Selfesta,' 'DRW3126,' 'Stamm 157/93'; and one cherry type 'Supersweet 100.' Each cultivar was grown, harvested and analyzed in three randomized replicates.

Experiment 2

In the second experiment, the concentration of the nutrient solution was varied. Increased concentration of nutrients can increase the soluble solids, sugar, acid and aroma volatile contents of the harvested fruits which in turn could improve sensory taste ratings (Stevens et al., 1979;

Adams, 1988; Holder and Christensen, 1988; Cornish, 1992; Soliman and Doss, 1992). But cultivar reactions did not always coincide (Caro et al., 1991). Therefore the longlife tomato cultivar 'Vanessa' and the round tomato type 'Counter' were cultivated in different concentrations of nutrient solutions, controlled by the electric conductivity (1, 3.5 and 6 dS/m). Each cultivar and treatment was grown in a separate row, replicated three times.

Sugar and Acid

The content of titratable acid was determined by potentiometric titration with 0.1 M NaOH, LMBG (1983). Sugar content was detected as glucose and fructose (Boehringer Mannheim, 1986) and summarized to reducing sugars.

Texture

Fruit firmness was measured using a deformation test. A force was applied to the calyx end of the fruit by a flat plate moving at a crosshead speed of 50 mm/min. Displacement at the force of 10 N was recorded and the pressure calculated using Equation (10.1):

$$P = \frac{F * 1000}{[(h/2)^2 - (h/2 - d/2)^2] * 2 * \pi} \qquad (10.1)$$

P = pressure firmness (kPa)
F = deformation force (N)
h = fruit height (mm)
d = displacement (mm)
π = Pi

Puncture force was recorded as maximum force (N) applied at the blossom end of the fruit using a probe with 3.2 mm in diameter and a crosshead speed of 50 mm/min.

Sensory Description

The Quantitative Descriptive Analysis (QDA) was conducted by a trained panel (Stone et al., 1974; Stone and Sidel, 1993) assessing 58 attributes including 'sweet note' and 'juicy' while chewing under defined conditions in single cabins. Scores were made on unstructured line scales. Anchor points were: 0 = not perceptible and 100 = strongly perceptable.

Consumer Acceptance

Acceptance tests with a representative target group of 98 housewives were implemented. The assessors judged the products' characteristics: first impression, appearance, smell, taste, aftertaste, mouthfeel in general and, as a special complement to the mouthfeel, both characteristics firmness whilst chewing and peel remaining in the mouth. As a final question the overall recommendability was asked. We used unstructured scales with the anchor points 0—*very unpleasant* and 100—*very pleasant*. For the characteristic firmness whilst chewing we used an unstructured ideal point scale with the anchor points: 0 = too soft, 50 = ideal, 100 = too firm.. The scores from ideal to too soft and from ideal to too firm were treated separately.

SWEET NOTE

Experiment 1

The amount of reducing sugars and titratable acids was significantly different in the tomato varieties. The amount of reducing sugars and titratable acid was not significantly related to the consumer acceptance ratings, but the "sweet" impression, which was detected by the members of the quantitative descriptive panel, clearly was related to the acceptance rating "recommendable." We had no explanation for this deviation from close relationship between sugar concentration and sweet descriptive impression (Table 10.3).

A principal-component analysis, which combined sensory and instrumental data, showed the close association between the descriptive attribute sweet impression and the consumer acceptance value for "recommendable." The descriptively assessed juiciness and the concentration of reducing sugars only showed a loose correlation with the first principal component (Figure 10.1).

Experiment 2

An increase of sugar and acid concentrations by plant nutrition with more concentrated nutrient solution could be demonstrated in Experi-

Table 10.3. Compounds and Descriptive Attributes of the Treatment That
Were Rated Lowest, Mean and Highest for "Recommendable"
by Consumers in Experiment 1.

| | Experiment 1 | | |
	Lowest	Mean	Highest
Recommendable	51.7	60.1	74.1
Red. sugars (g/100g)	3.0	2.9	2.5
Titr. acid (mg/100g)	478	560	532
Ratio sugar/acid	6.4	5.2	4.6
Descriptive sweet note	11.7	21.1	26.1
Descriptive: juicy	49.3	62.3	64.0

ment 2. But the highest sugar and acid values did not coincide with
higher intensities perceived by the descriptive panel. The consumer acceptance ratings for "recommendable" were lower at the highest EC levels (Table 10.4).

The principal component plot for this experiment also shows the dislike of peel fragments and fruit firmness being negatively correlated with
the first principal component (Figure 10.2). Concentrations of acids and

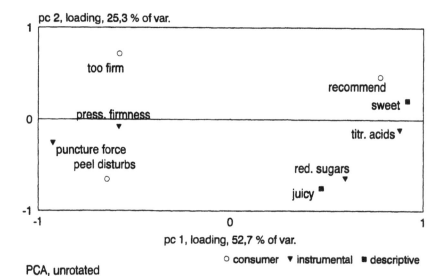

FIGURE 10.1 Principal-component analysis of sensory and instrumental data of tomatoes in the first experiment.

Table 10.4. *Consumers' Acceptance Ratings for "Recommendable" and Compounds and Descriptive Attributes of Two Cultivars under Three Nutrient Solution Regimes.*

	'Vanessa'			'Counter'		
	EC 1	EC 3.5	EC 6	EC 1	EC 3.5	EC 6
Recommendable	46.3 b	53.4 b	35.3 a	53.5 a	61.2 b	52.8 a
Red. sugars (g/100g)	2.90 a	3.16 b	3.49 c	2.68 a	3.08 b	3.22 b
Titr. acid (mg/100g)	416 a	529 b	590 c	416 a	472 b	526 c
Ratio sugar/acid	6.97	5.97	5.91	6.44	6.53	6.12
Descriptive sweet note	12.1 a	11.5 a	16.2 a	15.0 a	18.2 a	14.1 a
Descriptive: juicy	53.4 a	44.4 a	39.3 a	69.8 a	63.1 a	47.3 b

189

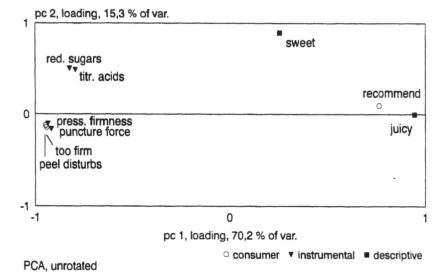

FIGURE 10.2 Principal-component analysis of sensory and instrumental data of tomatoes in the nutrient solution experiment.

sugars also were negatively correlated with the same factor. The consumer acceptance ratings for "recommendable" were positively correlated with the first component, as was the intensity of the descriptive juiciness. The descriptive sweet note clearly did not correlate the same way as sugar concentrations. The possible beneficial effect of more sugar and/or acid could thus not be detected, but the descriptive data for juicy show a relationship to the liking ratings.

The descriptive attribute sweet note was neither dependent on sugar content only nor on the juiciness, but on a combination of both. The combination of both may be the effective stimulus for the sweet perception, not the sugar concentration as such. A multiple-regression analysis of descriptive and analytic data in the combined data of the first and the second experiment showed that the sweet perception ratings of the descriptive panel were highly significant ($R^2 = 0,93$) and correlated to reducing sugars when combined with juiciness (Figure 10.3).

$$\text{sweet note} = -57.8 + 0.408 * \text{juiciness} \\ + 16.414 * \text{reducing sugars (g/100g)} \quad (10.2)$$

$$R^2 = 0.93$$

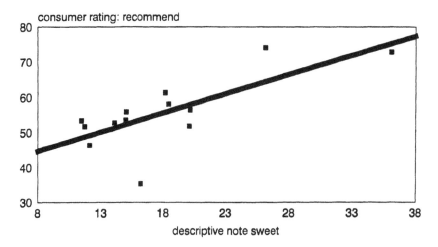

FIGURE 10.3 Acceptance vs. descriptive note sweet: recommendable = 36.03 + 1.09 * sweet note, $R = 0.73$.

Juiciness was significantly reduced by the treatments; perhaps the perception of the sugars is reduced in presence of low juiciness. The ratio reducing sugars to titratable acids was almost unchanged throughout the experiments and could therefore not be identified as contributing to sweet note or a consumer acceptance.

The combination of an instrumental method to estimate a juicy mouthfeel and the sugar concentration could work as a predictor for the sweet note of tomatoes, which contributes considerably to consumer acceptance. The sweet taste proved to have (within the analyzed range) a positive effect on consumer acceptance, which other contributors can counteract, like off-flavors or adverse texture properties as in the case of firmness.

FRUIT AND PEEL FIRMNESS

Closely related to the consumer dislike rating "too firm" and "peel disturbs" were the instrumental data for the pressure firmness and puncture force. This can be seen in Figure 10.1 and more clearly in Figure 10.2. In Experiment 2 the range of the instrumental values was much higher, counteracting acceptability (Table 10.5).

The recommendability of tomatoes was reduced when high puncture

Table 10.5. Texture Measurement Ranges and Consumer Dislike Values.

	Experiment 1			Experiment 2		
	min	mean	max	min	mean	max
Puncture force (N)	5.4	9.6	12.1	11.3	13.2	15.1
Pressure firmness (kPa)	30.4	35.0	42.0	30.2	41.0	53.9
Dislike firmness	0.0	1.6	11.2	0.0	11.5	29.4
Peel disturbs	30.9	39.6	47.2	42.9	50.3	56.5

forces were measured (Figure 10.4). The adverse effect of increasing firmness was consistent in both experiments. The dislike of soft fruits amounted only to low absolute values, and there was no negative correlation to the liking of the mouthfeel, and so on.

CONSUMER CLUSTERS

Auerswald et al. (1999) reported a tomato storage experiment, where initial fruit firmness was assessed ideal, too soft and too firm by one third of the consumers each. Postharvest softening led to significant differences in overall impression (liking) and liking of taste, dependent on

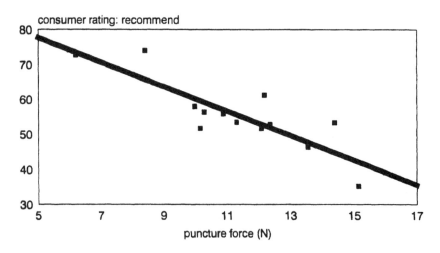

FIGURE 10.4 Acceptance vs. instrumental measurement: recommendable = 95.3 − 3.51 * puncture force, R = 0.84.

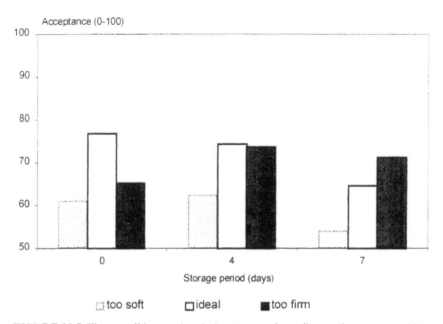

FIGURE 10.5 The overall impression during storage, depending on the consumers' initial firmness perception of the fresh fruits. Three groups were distinguished: those that did not like firm fruits, those that did not like soft fruits, and those perceiving fruits as *ideal*.

which of the three clusters the consumers belonged to (Figure 10.5, Figure 10.6). These differences were of the same magnitude, as the changes were caused by the storage period. An interpretation of the Hedonic relevance of the postharvest changes therefore is only possible against the background of the consumer segment.

Almost the same pattern of consumer evaluation was found in the case of flavor liking (Figure 10.6). Stored fruits were scored significantly higher than fresh fruits by those who felt fresh fruits were *too firm*. The similarity between the flavor liking and overall impression graphs suggests a combined pattern: Softer fruits exhibit taste qualities that are accepted in combination with the texture properties themselves. Those consumers who are disappointed by soft fruits are also disappointed by their taste and those consumers liking softer fruits find their taste more acceptable. A predictive model of the consumer liking of tomatoes should contain data for titratable acids, puncture force, reducing sugars and juiciness.

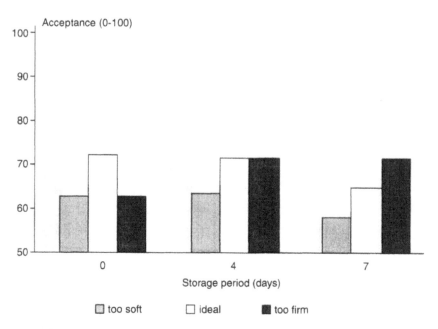

FIGURE 10.6 The approval of flavor during storage, depending on the consumers' initial firmness perception of the fresh fruits. Three groups were distinguished: those that did not like firm fruits, those that did not like soft fruits, and those perceiving fruits as *ideal.*

CONCLUSION

A stimulus may consist of a single chemical compound or physical property. Both should be able to be measured with the appropriate precision and accuracy. A stimulus may also consist of several components, which may or may not be analyzable by the human sensory system. In this case we cannot be sure that changes in one component directly translate into predictable sensations. They may change the pattern subjects will recognize, so we should investigate what contributes to one indivisible pattern and how their interactive mechanisms can be specified and quantified.

One consequence for fruit and vegetable production and handling means offering and controlling product concepts adapted to consumer segments, rather than relying on single or independent attributes. The knowledge of rejection points, where attributes dominate consumer acceptance ratings, as was found in case of the puncture force, or of the interaction of several descriptive and/or instrumental attributes, as in the case of sweet note perception of tomatoes, may facilitate the development of guidelines for production and quality control.

SUMMARY

The overall aim of bringing together instrumental data and consumer acceptance data is to become able to predict a consumers' Hedonic reaction to perceivable attributes using objective, measurable data. The objective of this research for horticultural products is, besides the purely scientific, to develop guidelines for production and offering criteria for quality assurance and management.

A consumer survey showed that the important quality criteria applied to fruits and vegetables can be grouped into four major categories. The first category includes appearance, color and texture, which can be readily determined visually or by touch. Therefore, they can be used as criteria throughout the production and distribution system. The second category includes flavor and mouthfeel. In both segments, the criteria are closely related to physical and chemical attributes and are perceivable by the human senses during consumption and by analytic instruments.

Experiments with different tomato varieties were conducted to investigate the relationship between measured, instrumental data (puncture force, pressure firmness, reducing sugars, titratable acid) and consumer acceptance ratings. The amount of reducing sugars was not directly correlated with consumer ratings, as was the sweet note, determined by Quantitative Descriptive Analysis (QDA). Multiple regression showed that reducing sugars and juiciness (again determined by quantitative descriptive analysis) seemed to interfere, combined explaining most of the variation of the note "sweet" by QDA ($R^2 = 0.93$), thus having high potential as a predictor for consumer acceptance.

Puncture force readings, associated with disturbing peel fragments, consistently led to increased dislike ratings as shown by PCA and linear regression.

REFERENCES

Adams, P. 1988. Some response of tomatoes grown in NFT to sodium chloride. *ISOSC Proceedings,* Flevohof, NL, 7: 59–70.

Amerine, M. A., Pangborn, R. M. and Roessler, E. B. 1965. Physical and chemical tests related to sensory properties of foods, in *Principles of Sensory Eval-*

uation of Food, M. L. Anson, C. O. Chichester, E. M. Mrak, G. F. Stewart, eds., Academic Press, New York, pp. 494–518.

Andani, Z. and MacFie, H. J. H. 1999. Consumer preference, in *Fruit and Vegetable Quality: An Integrated View,* R. L. Shewfelt and B. Brückner, eds., this volume.

Auerswald, H., Peters, P., Brückner, B., Krumbein, A., and Kuchenbuch, R. 1999. Sensory analysis and instrumental measurements of short-term stored tomatoes (*Lycopersicon esculentum* Mill.). *Postharvest Biology and Technology,* 15: 323–334.

Boehringer Mannheim GmbH, Biochemica (Mannheim, Deutschland), 1986. Methoden der biochemischen Analytik und Lebensmittelanalytik, pp. 37–38.

Booth, D. A. 1990. Designing products for individual customers, Ch. 8, in *Psychological Basis of Sensory Evaluation,* McBride, R. L. and H. J. H. McFie, eds., Elsevier Science, Essex, U.K., pp. 163–193.

Bourne, M. C. 1983. Correlating instrumental measurements with sensory evaluation of texture, Ch. 3.3, in *Quality in Food and Beverages,* A. A. Williams and R. K. Atkin, eds., Ellis Horwood, England, pp. 155–172.

Brown, W. E., Dauchel, C., and Wakeling, I. 1996. Influence of chewing efficiency on texture and flavor perceptions of food. *Journal of Texture Studies,* 27: 433–450.

Cardello, A. V. and Sawyer, F. M. 1992. Effects of disconfirmed consumer expectations on food acceptability. *Journal of Sensory Studies,* 7: 253–277.

Caro, M., Cruz, V., Cuartero, J., Estan, M. T., and Bolarin, M. C. 1991. Salinity tolerance of normal-fruited and cherry tomato cultivars. *Plant and Soil,* 136(2): 249–255.

Conner, M. T. 1994. An individualised psychological approach to measuring influences on consumer preferences, Ch. 7, in *Measurement of Food Preferences,* H. J. H. MacFie and D. M. D. Thomson, eds., Chapman & Hall, London, pp. 167–201.

Cornish, P. S. 1992. Use of high electrical conductivity of nutrient solution to improve the quality of salad tomatoes (*Lycopersicon esculentum*) grown in hydroponic culture. *Australian Journal Exp. Agric.,* 32: 513–520.

Florkowski, W. J., Brückner, B., and Schonhof, I. 1996a. Nutritional attributes as reasons for German consumer fruit consumption. *J. Agric. Applied Economics,* 28(1): 235–236.

Florkowski, W. J., Brückner, B., and Schonhof, I. 1996b. Importance of produce freshness to European consumers: evidence from Berlin, Germany. *Acta Horticulturae,* 429: 135–140.

Grunert, G. G., Larsen, H. H., Madsen, T. K., and Baadsgaard, A. (eds.) 1996. *Market Orientation in Food and Agriculture,* Kluwer Academic Publishers, Boston.

Grunert, K. G. 1995. Food Quality: A means-end perspective. *Food Quality and Preference,* 6: 171–176.

Holder, R. and Christensen, H. 1988. The effect of electrical conductivity on the growth and composition of cherry tomatoes grown on rockwool. *ISOSC Proceedings*, Flevohof, NL, 7: 213–228.

Kapsalis, J. G. and Moskowitz, H. R. 1978. Views on relating instrumental tests to sensory assessment of food texture. Applications to product development and improvement. *Journal of Texture Studies*, 9: 371–393.

Kroeze, J. H. A. 1990. The perception of complex taste stimuli, Ch. 3, in *Psychological Basis of Sensory Evaluation*, R. L. McBride and H. J. H. McFie, eds., Elsevier Science, Essex, U.K., pp. 41–68.

Land, D. G. 1983. What is sensory quality? Ch. 1.1, in *Quality in Food and Beverages*, A. A. Williams and R. K. Atkin, eds., Ellis Horwood, England, pp. 15–29.

Lawless, H. 1990. Applications of experimental psychology in sensory evaluation, Ch. 4, in *Psychological Basis of Sensory Evaluation*, R. L. McBride and H. J. H. McFie, eds., Elsevier Science, Essex, U.K., pp. 69–92.

Lennernäs, M., Fjellström, C., Becker, W., Giachetti, I., Schmitt, A., Remaut de Winter, A. M. and Kearny, M. 1997. Influences on food choice perceived to be important by nationally representative samples of adults in the European Union. *European Journal of Clinical Nutrition*, 51(2): 8–15.

LMBG (Amtliche Sammlung von Untersuchungsverfahren nach § 35 LMBG), 1983. Bestimmung des Gesamtsäuregehaltes von Tomatenmark (potentiometrische Methode) L 26. 11. 03–4.

Malundo, T. M. M., Shewfelt, R. L., and Scott, J. W. 1995. Flavor quality of fresh tomato (*Lycopersicon esculentum Mill.*) as affected by sugar and acid levels. *Postharvest Biology and Technology*, 6: 103–110.

McBride, R. L. and Anderson, N. H. 1990. Integration psychophysics, Ch. 5, in *Psychological Basis of Sensory Evaluation*, R. L. McBride and H. J. H. McFie, eds., Elsevier Science, Essex, U.K., pp. 93–117.

McBride, R. L. 1990. Three generations of sensory evaluation, Ch. 9, in *Psychological Basis of Sensory Evaluation*, R. L. McBride and H. J. H. McFie, eds., Elsevier Science, Essex, U.K., pp. 195–206.

Meilgaard, M., Civille, G. V., and Carr, B. C. 1991. Affective tests: Tests and in-house panel acceptance tests, Ch. 10, in *Sensory Evaluation Techiques*, CRC Press LLC, Boca Raton, FL, p. 213.

Moskowitz, H. R. and Krieger, B. 1993. What sensory characteristics drive quality? An assessment of individual differences. *Journal of Sensory Studies*, 8: 271–282.

Moskowitz, H. R., Krieger, B. 1995: The contribution of sensory liking to overall liking: an analysis of six food categories. *Food Quality and Preference*, 6, 83–90.

Moskowitz, H. R., Krieger, B., and Cohen, D. 1994. Meeting the Food Market Challenge: Creating Winning Food Concepts for Consumers in a "Real Time" Mode.

Presented at Health and Pleasure at the Table. A Symposium held at the University of Montreal, Montreal, Canada, May 1994.

Norwich, K. H. 1991. Towards the unification of the laws of sensation: Some food for thought, Ch. 5, in *Sensory Science. Theory and applications in Foods,* H. T. Lawless and B. T. Klein, eds., Marcel Dekker, Inc., New York, pp. 151–183.

Oude Ophuis, P. A. M. and van Trijp, H. C. M. 1995. Perceived quality: a market driven and consumer oriented approach. *Food Qual. Pref.,* 6: 177–183.

Pangborn, R. 1981. Individuality in responses to sensory stimuli, in *Criteria of Food Acceptance,* J. Solms and R. L. Hall, eds., Forster Verlag AG, Zürich, pp. 177–217.

Schutz, H. G. 1990. Measuring the relative importance of sensory attributes and analytical measurements of consumer acceptance. *Acta Horticulturae,* 259: 173–174.

Schutz, H. G. 1993. Measuring consumer acceptance of flavors, Ch. 5, in *Flavor Measurement,* C. T. Ho and C. H. Manley, eds., Marcel Dekker, New York, pp. 95–112.

Solimann, M. S. and Doss, M. 1992. Salinity and mineral nutrition effects on growth and accumulation of organic and inorganic ions in two cultivated tomato varieties. *J. Plant Nutrition,* 15(12): 2789–2799.

Sonneveld, C. and Straver, A. 1988. Voedingsoplossing voor groenten en blomen in water of substraten. Consulentenschap voor de Tuinbouw. Proefstation voor Tuinbouw onder Glas. *Naaldwijk,* 11: 36.

Stevens, M. A., Kader, A. A., and Albright-Holton, M. 1979. Potential for increasing tomato flavor via increased sugar and acid content. *J. Amer. Hort. Sci.,* 104(1): 40–42.

Stone, H. and Sidel, J. L. 1993. Descriptive analysis, in *Sensory Evaluation Practices,* H. Stone and J. L. Sidel, eds., Academic Press, Orlando, FL.

Stone, H., Sidel, J. L., Joel, O. S., Woolsey, A., and Singleton, R. C. 1974. Sensory evaluation by quantitative descriptive analysis. *Food Technol.,* 28. 24–34.

Szczesniak, A. S. and Kahn, E. L. 1971. Consumer awareness of and attitudes to food texture. I. Adults. *Journal of Texture Studies,* 2: 280–295.

Thompson, D. M. H. and MacFie, H. J. H. 1983. Is there an alternative to descriptive sensory assessment? Ch. 2.4, in *Sensory Quality in Food and Beverages,* A. A. Williams and R. K. Atkin, eds., Ellis Horwood, England, pp. 96–107.

CHAPTER 11

House of Quality—An Integrated View of Fruit and Vegetable Quality

ANNE C. BECH

INTRODUCTION

HOUSE of Quality is the most important element of Quality Function Deployment (QFD)—the first set of matrices—in which customer or consumer needs and wants are translated into quality characteristics (measurable technical attributes) (Hauser and Clausing, 1988). By the use of QFD and the House of Quality, the objective for product development is defined in a terminology that is understandable by the developers and based directly on customer needs. In order to understand how House of Quality can be used, we have to consider the goal of product development and what QFD is.

The long-term goal of product development for seed producers, fruit and vegetable growers, food companies, retailers and other companies, is to obtain long-term competitive advantages. According to Day and Wensley (1988), this goal can be achieved in two ways. Manufacturers can either make products that consumers perceive to be better than competing products, or they can make them at relatively lower cost than their competitors. Both of these aspects are included in QFD in the mentioned order. In the case of the introduction of a new product, it is crucial that the consumer perceives it to have a greater value than existing products.

QFD is an abstract concept that can be described as a market-oriented (proactive), integrated product development process. The starting point is in the consumer's (the customer's) needs, and the aim is to build qual-

199

ity that the consumer (the customer) perceives as an advantage into new products, expressed as *listen to the voice of the customer*. The term "customer" is general and refers to customers in the whole chain from the end user (the consumer), to the retailer, all the middlemen, the fruit or vegetable grower, the seed supplier and, finally, the seed developer. In this chapter the term "consumer" is used to distinguish the ultimate consumer from other customers.

QFD originated in the Japanese shipbuilding industry in 1972 (Akao, 1990) and is closely related to the Japanese management philosophy Total Quality Management. QFD spread from the shipbuilding industry to the electronics and car industries in Japan, and later to a number of other industries throughout the world. According to Hofmeister (1991), it was adopted by the American food industry in 1987. QFD has been described by Akao (1990), Mizuno and Akao (1994), Day (1993), Cohen (1995), Urban and Hauser (1993), and Hofmeister (1991). Since product development is usually a company-specific undertaking surrounded by a high wall of secrecy, actual examples of the use of QFD in the development of food products are few and far between (Bech et al., 1997a; Bech et al., 1997b). Descriptions of and recommendations for the use of QFD as well as examples to illustrate of the method are much more frequent (e.g., Hofmeister, 1991; Charteris, 1993; Pedi and Mosta, 1993; Swackhamer, 1995; Dalen, 1996).

QFD covers quality, technology, cost and reliability deployment (Akao, 1990) and represents a relatively new approach to product development in the food industry. First, it creates a common basis for the development of new products; namely, an understanding of consumer needs and wants, which are explicitly identified in the initial phase of the product development process. Second, QFD provides a framework for organizing information from different parts of the firm throughout the product development process. Third, QFD links information from consumers to the firm's internal production processes and quality assurance systems. All information is collected in tables (matrices), which can be seen consecutively, such that columns in the first table become rows in the next, and so on. Finally, QFD is flexible, it can and should be adapted to the individual project, team or firm (Mizuno and Akao, 1994). The use of the QFD model for product development means that consumers' needs are central throughout the product development process, plus products are designed specifically to fulfil important needs seen from the consumers' point of view.

This differs from the more traditional product development process (see, e.g., Earle, 1997), which often begins with idea generation, either on the basis of market information the firm already possesses or market surveys. This phase is seen as a creative phase, the aim of which is to generate ideas for concepts and products, which are subsequently tested in various consumer tests. Decisions are therefore related to the concept that is thought to be best suited to satisfying consumers' preferences. Whether the end product is successful or not, therefore, depends to a large extent on how good firms are at developing concrete products and concepts. For the QFD team, it is more a question of finding good solutions to consumers' needs. In other words, firms' technological or agricultural competencies are directed towards consumers' needs.

The purpose of this chapter is to introduce QFD and the House of Quality as an integrated quality concept and describe the two central elements of QFD, food quality and consumer needs, with special attention to the translation processes for fruit and vegetables. Finally, an example of translation of consumer needs for good sensory quality into measurable sensory attributes is given.

With fruit and vegetables in mind, obvious areas for the use of House of Quality and the integrated view of quality could be the development of completely new fruits and vegetables; a new variety; special sizes, cuts, assortments; packaging; distribution systems; or the establishment of fruit and vegetable departments in supermarkets, all in accordance with consumer (and customer) needs and wants and within the technical possibilities available. It is, however, obvious that the use of House of Quality in this area may be less technically oriented than within other types of products. This part is discussed in the section Translation of Needs.

FOOD QUALITY

Food quality has been the subject of endless debate, and this has inevitably given rise to a proliferation of different terms and definitions recently discussed by, e.g., Shewfelt et al. (1997). What is needed for the product development and quality control of food, however, is a versatile and dynamic quality concept that can also be related directly to the goal of product development, i.e., the creation of long-term competitive advantages, perceived advantages in consumers' and other customers' eyes.

From a strategic point of view, there is wide agreement on the following quality definition for the goal of product development: "products and services that satisfy or exceed the customer's requirements" (Surak and McAnelly, 1992, p. 80), while, according to Oliver (1997, p. 98), it is "exceed the customer's expectations."

This is also called *subjective* quality, the quality which the customer or consumer perceives in the product. Quality (value) is attributed to the product because, from the consumer's point of view, it has some advantages and satisfies some wants. This differs from *objective* quality, which constitutes the total measurable or documentable attributes of a product (see Grunert et al., 1996), or, as Oliver (1997, p. 151) puts it, "What the consumer gets" in relation to "What the product has." This distinction was already made by Shewhart (1931) and maintained in the later ISO definition of quality.

According to Oliver (1997), there is a general view that it is sufficient to fulfil the customer's expectations as suggested by Feigenbaum (1961). However, Oliver (1997) emphasizes that firms that aim to go further will do better at the expense of firms that merely aim at satisfying customers' expectations. He recommends that, with regard to maximizing customer satisfaction, firms should seek to create high expectations in the consumer but that it should still be possible for the product to exceed these expectations.

The quality assessment of food products prior to purchase is based on the consumer's *expectations* that the product will satisfy a need, whereas the overall quality assessment is also based on the consumer's *experiences* of the product. A general model has been developed by Grunert et al. (1996), and a revised version is illustrated in Figure 11.1.

The model has been revised and simplified in several aspects: (1) the dependent variable on the right side is generalized to an "overall quality" evaluation instead of "experienced purchase motive fulfillment" and "future purchases" and the quality formation process is included (Poulsen et al., 1996); (2) the term "cost" includes price and the case where price acts as an extrinsic cue has been illustrated by a dotted arrow; (3) "purchase decision" replaces "intention to buy" and "expected purchase fulfillment"; (4) "perceived cost cues" covers both "perceived cost cues" and "perceived costs" and (5) the model is horizontal with objective quality on the left side.

I consider the model general as earlier research contributions are included: the distinction between extrinsic and intrinsic quality cues (Olson and Jacoby, 1972), the distinction between objective and subjective quality (Shewhart, 1931), the distinction between expected and experienced

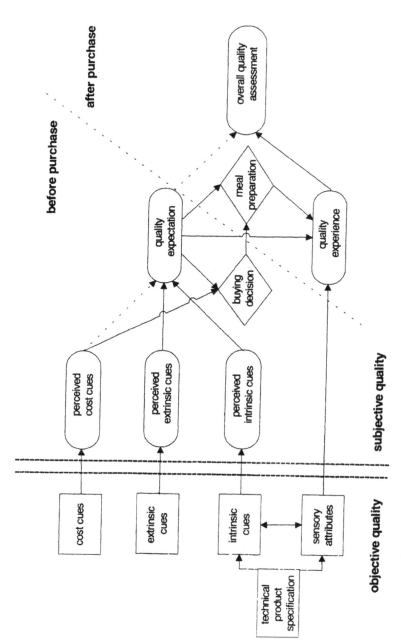

FIGURE 11.1 A revised and simplified Total Food Quality model (after Gunert et al., 1996).

before purchase   after purchase

overall quality assessment

meal preparation

quality expectation

buying decision

quality experience

perceived cost cues

perceived extrinsic cues

perceived intrinsic cues

cost cues

extrinsic cues

intrinsic cues

sensory attributes

technical product specification

subjective quality

objective quality

203

quality (Steenkamp, 1989) and the quality formation process (Poulsen et al., 1996). Furthermore, the expectation-disconfirmation framework recently discussed by Oliver (1997) is also related to the model.

As we aim at meeting or exceeding consumers' expectation a discussion of the relationship between the Total Food Quality Model and the expectation-disconfirmation model follows.

In the expectation-disconfirmation framework the overall satisfaction/dissatisfaction is a function of the expectations and their disconfirmation. In the *assimilation process* we have a direct effect of the expectations and the experienced quality approaches expectations, even though a difference is perceived. This is the main process in the case of minor differences between expected and experienced quality. Alternatively, the *contrast process* may occur. In this case the expectations are thoroughly disconfirmed. High expectations that have not been fulfilled will result in an even lower evaluation of the product than if expectations had been lower, and vice versa. Hence exceeding consumers' expectations corresponds to a positive contrast effect. The assimilation process has been illustrated by numerous examples but not the contrast effect (Oliver, 1997).

Five submodels of the Total Food Quality Model are illustrated in Poulsen et al. (1996). The *pure assimilation process* corresponds to the expectation-only sub-model, while the *pure contrast process* corresponds to an experience-only submodel. Furthermore, three models include the effects of both expectations and experience, the full model (as illustrated in Figure 11.1), the intervening model (as illustrated in Figure 11.1 without the dotted arrow) and the additive model. Exceeding consumer expectations corresponds to one of these three submodels with positive coefficients.

The constructs in Total Food Quality Model may be considered as latent and should each be represented by measurable manifest variables at the attribute level. This again gives rise to various model combinations, which, however, are outside the scope of this chapter. Nevertheless, it is important to stress that a model at the attribute level implies that the consumer's expectations of some attributes can be fulfilled at the same time as others are experienced as disappointing. A simple example is a product whose taste satisfies the customer, but where the packaging does not and Cardello (1994) points out that the consumer can have expectations of both specific sensory experiences and of how the product in general will live up to expectations. Examples of various subsets of the Total Food Quality model are given by Grunert et al. (1996). The model may be estimated by a structural equation system, one example is Poulsen et al. (1996).

Finally, the Total Food Quality model can serve as an overall framework for the analysis of consumers' quality perception and its relation to the design of food products (Grunert et al., 1996). This theoretical approach is particularly relevant for the future work with the House of Quality where each of the constructs in the model are operationalized in terms of manifest variables of both subjective and objective quality.

HOUSE OF QUALITY

In order to concretize what QFD is, we can take a look at the type of information collected in the House of Quality in Figure 11.2, the first matrix in QFD (Hauser and Clausing, 1988). This consists of a number of tables organized in such a way as to resemble a house. It is essential to understand how a House of Quality is formed and "read." In many respects, it is like drawing and reading a map.

The House of Quality consists of two dimensions, the horizontal consumer (customer) dimension and the vertical company (technical) dimension. The house is read and filled in starting at the far left with the consumer dimension, which consists of three parts: the consumer needs and wants (whats), the importance of these and the consumer perception (ratings) of the company's own product and main competing products at the right-hand side. The consumer needs are usually identified in qualitative consumer studies, for instance, focus groups. For food, including fruit and vegetables, the voice is closely related to the senses of the consumer, since one of the most common requirements is good sensory quality. The consumer needs include attributes that they find important for good sensory quality for specific foods. The importance of the needs is measured or estimated in the following quantitative study in which consumers' perceptions of the sensory quality are measured for each of the identified attributes as well as overall acceptance. The sensory analysis involving consumers is treated as an integrated part of the market research and the quantitative study includes measurement of consumer behavior and attitudes as well as demographic questions.

The technical dimension consists of the quality characteristics (measurable or documentable attributes) corresponding to the consumer needs. At the bottom of the House of Quality, we find the technical assessments on these attributes.

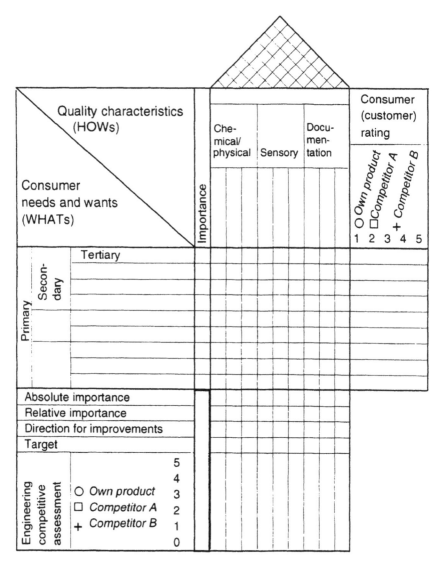

FIGURE 11.2 Elements in the house of quality.

The relationship matrix, the central part of the House of Quality, is the place where the consumer and company dimensions meet. This is usually formed by the responsible marketing and R&D teams, which discuss the relevance of various measuring methods to the identified consumer needs. Alternatively, quantitative techniques can be used for the modeling consumers' perception of a range of products; examples include Gustafson (1993), Bech et al. (1997a) and Bech et al. (1997b).

The most important task is to analyze the information in the House of Quality. Strategic possibilities exist where consumers have wants that are not satisfied by the products provided so far, or where the consumers perceive the actual products as being of higher quality. Where the quality is perceived as being lower, the direction for improvement can be identified by comparisons with the technical measurements.

Finally, it is possible to show the connections between the measurements in the "roof." Various PC programs have been developed to support the use of QFD in product development, e.g., QFD Designer, QFD Capture and recently QFD 2000. Microsoft's Excel spreadsheet is also widely used, especially for larger projects.

CONSUMER NEEDS

From a material point of view, the standard of living of consumers in the Western world is high enough to meet most basic needs. This also applies to food. More or less everybody can at any time buy and eat enough food to satisfy the most basic of needs—hunger. However, food also has another role than the mere satisfaction of hunger, namely, to contribute to the quality of life, to happiness and pleasure. These two situations can be characterized as needs and wants, respectively. The dividing line between needs and wants is fluid, however. As wants are satisfied, the boundary moves, so that what was once seen as wants become regarded as an absolute necessity. Foods meet social needs (e.g., meals with friends) and contribute to a person's identity (e.g., ecological food), too. The phrase "you are what you eat" sums this up nicely. Food can thus help satisfy needs at all levels, e.g., related to Maslow's hierarchy of needs.

Kano et al. (1982) (see Mazur, 1987) distinguish between three different kinds of product attributes according to how they contribute to satisfaction. Some product attributes contribute to the must-be (basic) quality perceived by the customer, others contribute to a proportional increased satisfaction, while a third type of quality attributes (exiting/attractive) helps exceed the customer's expectations. It is necessary that the must-be quality attributes are included before the desired positive effects of "exiting" quality attributes can be achieved. Oliver (1997) directly links needs fulfillment with satisfaction and uses three categories of needs, which, if fulfilled, result in different degrees of satisfaction.

The categories are then related to the product attributes. Oliver (1997) distinguishes between: bivalent satisfiers—the upward and downward translatable attributes that can cause both satisfaction and dissatisfaction (proportional quality in Kano's model); monovalent dissatisfiers—essential but unprocessed attributes only capable of causing dissatisfaction when flawed (must-be quality in Kano's model); and monovalent satisfiers—psychological extras processed at a higher level of the needs hierarchy (exiting quality in Kano's model).

The Kano approach illustrates that it is difficult to achieve customer satisfaction and his approach covers aspects of both compensatory and noncompensatory models. So far, the Kano approach has not been modeled and the general second-order model (e.g., used by Sloof et al., 1996) is not sufficient either.

IDENTIFICATION OF CONSUMER NEEDS

In general, consumer needs should be identified by means of qualitative surveys. In QFD terminology, this is called *listen to the customer's voice.* Various methods can be used, but the most common are personal interviews and focus groups (Griffin and Hauser, 1993). Laddering interviews can also be used (Oliver, 1997), but usually result in an identification of needs at higher levels in the hierarchy (Griffin and Hauser, 1993) because they only include statements mentioned by a certain number of respondents.

For QFD purposes, needs are identified directly from consumers' statements. It is difficult to say precisely how many interviews are necessary to identify consumer needs satisfactorily, and only one study has been carried out in the area. Griffin and Hauser (1993) show that about two-thirds of consumer needs can be identified from four personal interviews or two focus groups, and that this rises to 80% with two extra personal interviews or one focus group. The value of additional interviews declines, and their study does not show an expected effect of group dynamics in the focus groups. They recommend 20–30 personal interviews, while others are more realistic and recommend 10–20 (Day, 1993). According to Griffin and Hauser (1993), it is important that the results are analyzed by persons with different backgrounds.

Commenting on Kano's model, Day (1993) says that basic needs (must-be quality) are rarely mentioned in qualitative studies; it is the functional attributes that are expected and usually fulfilled. Unless the customer has experienced the absence of these attributes, they are not mentioned. Instead, proportional quality is what consumers talk about.

Exiting quality, which is more than consumers expect, and thus unconscious, is seldom mentioned. It requires a creative leap based on customers' comments or observations of customer behavior. Generally, useful methods are those that do not necessitate verbalization, e.g., various mapping techniques based on preference measurement or one-on-one comparisons. However, this presupposes subsequent interpretation of the underlying latent attributes. Other nonverbal techniques, such as behavioral studies or the use of other stimuli, such as pictures or products, are also a possibility.

The need analysis is followed by a structuring of the needs in a hierarchy, which, according to Oliver (1997), does not have to follow a strict model. In QFD terminology, these are called *primary strategic needs, secondary tactical needs,* or *tertiary operational* needs. Needs can be structured by means of an affinity diagram (whereby needs are classified into groups, so that the needs in one group are comparable and different from needs in other groups). This can be done by a team responsible for product development or based on consumers' classification, which, according to Griffin and Hauser (1993), best reflects consumer needs.

Next, the focus shifts to how the different needs or needs-related attributes might influence satisfaction with the product. Here, Oliver (1997) stresses the importance of identifying needs in all three categories.

EXAMPLES OF CONSUMER NEEDS

In order to illustrate what needs are in relation to food, I will refer to several analyses and give some general examples related to Kano's model (Table 11.1). The needs depend on both the product, the individual consumer (segments), and the context in which the product is used. These needs should therefore be seen solely as examples of needs.

At the upper end of the hierarchy, needs are identified as: self-confidence and self-esteem (Gutman—see Oliver, 1997), the family's quality of life, good health and long life (Sørensen et al., 1996; Bech-Larsen et al., 1996). Further down the hierarchy, needs are related to health, pleasure from eating, many ways of cooking (fish), easier-to-make delicious dishes (fish), and good cooking results (Sørensen et al., 1986; Bech-Larsen et al., 1996). All the above examples also show that consumers want products that taste good.

Recent qualitative research in Denmark, Sweden and Germany (Skytte et al., 1997, unpublished results) shows that consumers perceive fruit

*Table 11.1. Categorization of Different Attributes
in Relation to Kano's Model*

Kano's Model	Oliver's Definition	Product Attributes
Exiting quality	Monovalent satisfier	Ecological
		Local fruit and vegetables
Proportional quality	Bivalent satisfier	New varieties
		Packaging
		Convenience
Must-be quality	Monovalent dissatisfier	Good taste
		No health risks

and vegetables as an essential part of a proper meal, as being suitable for weight reduction, as giving variation, as giving appealing colors to the dish, and that consumers want more fruit and vegetables. Consumers express needs for natural, fresh and tasty products that are easier to clean, peel and prepare. Besides, consumers ask for trustworthy information, not just different kinds of quality labels, and they prefer locally produced products. At the same time, they are very concerned about the loss of vitamins during processing, the use of packaging and packaging materials, gen-technology, pesticides, and irradiation. Obviously, development and marketing of new products based on these consumer needs require careful use of technology and packaging. In the development process quantitative consumer research is needed to ensure that the new product meet or, even better, exceed consumers' expectations.

It can be difficult to classify attributes in relation to Kano's model, but taking Oliver's descriptions into consideration helps. The extent to which a concrete attribute gives rise to exiting or proportional quality also depends on time. If it is possible to copy the attribute concerned, it will often end up being proportional or even must-be, while a form of protection of or relation to a brand helps create long-term positional advantages.

I have chosen to place the ecological and local attributes in the category "exiting quality," which denotes something extra compared with other products. At least at the Danish market, the organic products are accompanied by a quality label that gives consumers confidence and security about the production method; information on country of origin should be available and cannot be copied. It is also more likely that fruit and vegetables produced locally satisfy the expressed consumer need for fresh products. I regard good taste as a must-be attribute of food. Consumers are hardly likely to want food products that do not taste good, so this is also a necessary attribute of ecological products. That food has no health risks is also a must-be attribute, which should be guaranteed via legislation and production processes.

TRANSLATION OF CONSUMER NEEDS

After the identification and structuring of needs follows the translation where the QFD team suggests quality characteristics (measurable or documentable attributes, part of the objective product quality) that correspond to the identified needs. In the case of food products, it can be both technical and sensory attributes (Bech et al., 1994). As already mentioned, development of fruit and vegetables may be less technically oriented than more traditional food production. Many quality characteristics are based on qualitative or semi-objective judgements instead of actual measurements (e.g., the assessment of ripeness and robustness). A short and precise description of the quality criteria may be useful to support the assessments.

The general approach is described in the discussion of the House of Quality. With a focus on fruit and vegetables, it is important to bear in mind that the objective quality, when the consumer gets access to the product in a purchase or consumption context, is a result of the total process from seed development, growing, handling, storage and distribution—corresponding to the often many participants in the chain.

Traditionally, each participant pays most attention to the recent and/or the most powerful participant in the chain. This sequential approach may result in suboptimization in the chain, both with regard to product quality and profit. Alternatively, a parallel QFD approach covers all participants in the chain, as illustrated in Figure 11.3. This approach has many similarities to supply chain management where product flow and information flow play important roles (see, e.g., Gattorna and Walters, 1996). However, one major difference is that the QFD approach also considers product changes that are overlooked by the supply chain management approach. For fruit and vegetables, product changes may be either improvements (e.g., controlled ripening) or deterioration (e.g., overripening or spoilage). An attempt to model the quality changes is suggested by Sloof et al. (1996).

The figure illustrates five parts in the chain from seed producer (S), grower (G), wholesaler (W), and retailer (R) to the consumer (C). The

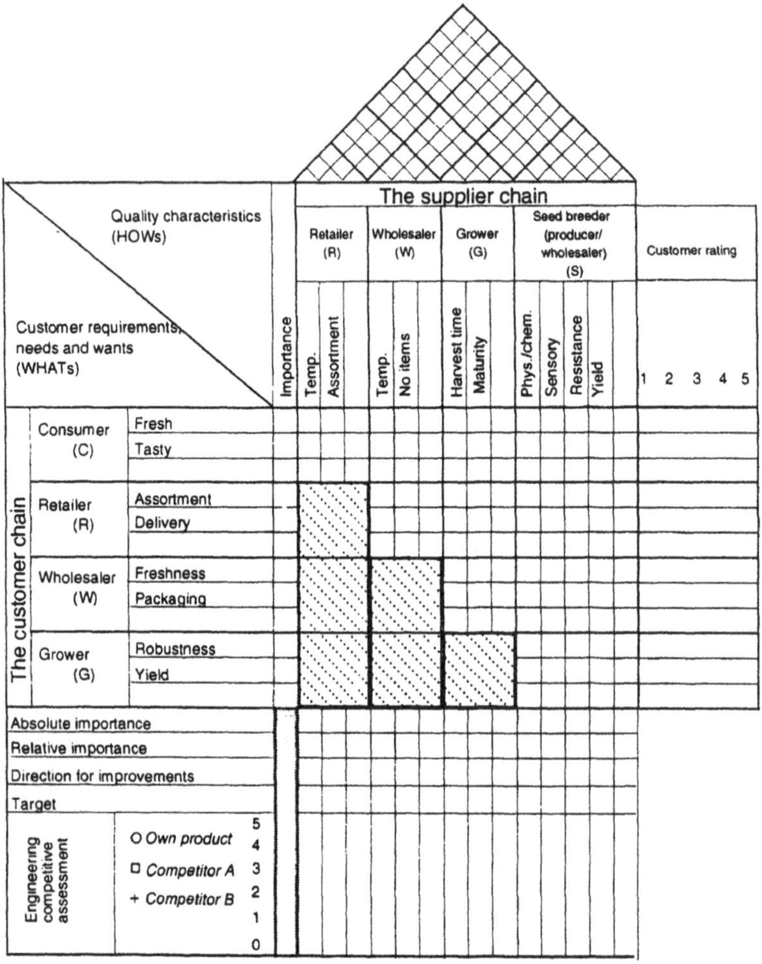

FIGURE 11.3 House of Quality for the chain from seed developer to consumer.

figure could be further adapted to both more complex and simpler situations. The complex case requires an expansion of the house where, e.g., some of the parts cover more participants, e.g., seed producer (S) versus seed developer, seed grower, seed supplier, while the more simple case is a shrinkage of the house, e.g., the grower could have a farm shop.

Horizontal: At the left side of the house we find the needs and requirements from all the customers in the chain with the consumer needs at the upper-left corner and the growers at the lower-left corner.

Vertical: From the left side, we find all the quality characteristics used in the supplier chain from the retailer ending with the seed breeder/ developer; in other words, how each of the participants uses different attributes for defining and controlling the quality of the fruit and vegetables.

The central part: In this example the relationship matrix consists of 16 different matrices; I assume that the six matrices in the lower-left part are irrelevant as only very powerful suppliers are able to impose conditions on the succeeding partners in the distribution chain. Reading the rest of the central matrix from the left, the consumers' needs and wants have to be considered by the retailer (R), the wholesaler (W), the grower (G) and the seed supplier (S) in order to maximize consumer satisfaction. In the same manner, R's requirements have to be considered by both W, G and S, and W's by the G and S. Finally, G, of course, imposes additional requirements on S. Reading the central matrix from the top, the four parts represent each part in the distribution chain and the total of needs, wants and requirements that they have to consider and the quality characteristics they may use to match the different requirements. The right part of the figure is unique for each participant as they all have different customers and competitors; e.g., S would compare C's perception of different varieties while R would compare C's perception of fruit and vegetables in different retail stores.

The roof: We find several interesting relationships here. The first level (the four triangles) represents each of the relationships between the quality parameters used by each participant in the chain. The second level (above the triangles) represents the relationship between the quality attributes used at customer and supplier level. The third level represents the relationship between the quality attributes for two participants with a part in between. Finally, the fourth level represents the relationship between the quality parameters used by R and the first part in the chain, S. The roof could be filled in with positive and negative relationships, leaving empty cells where there are no relationships.

If we assumed that R had a perfect knowledge of the consumer needs and perfect translations to attributes in all parts of the chain, the sequential approach would be appropriate and we would only need the four matrices C-R, R-W, W-G, G-S in the central part. In this case the decisions made by others than R are based solely on indirect information on consumer needs—an approach with high risk of failure in the decision process.

On the left side of the roof, reading from the bottom to the top, we find the sequence of relationships as if all participants listened to the most powerful part in the chain, in this case the retailer (R), and with ignorance of the direct study of consumer needs. On the right side of the roof, reading from the bottom to the top, we find the case as if the producer (or seed producer) pushes his quality characteristics on the market and the others have to set their quality characteristics accordingly.

Using the fully integrated House of Quality approach represents the opposite alternative, e.g., S would study all the requirements of C, R, W and G, decisions are consequently based on direct information, an approach that could be both time-consuming and resource demanding. Finally, as a more realistic approach the integrated House of Quality could be used to decide where direct information is needed and where indirect information is sufficient. In this connection both the number of translation processes and the source of information are critical factors.

From a research perspective, many of the relationships in the roof are well known; e.g., research on storage conditions and postharvest treatment related to different cultures and varieties.

In Figure 11.3, we have the translation of customer requirements into quality characteristics used by the different participants in the whole chain. The following phases are individual for each of the participants and include how to produce with attention to the cost and the safety (foodborne disease, toxins, etc.) and reliability (shelf life, off-taste, etc.) of the different alternatives (Mazur, 1994). In QFD terminology, the phases are called quality, technology, cost and reliability deployment (Akao, 1990). In Figure 11.4, the four phases (for the grower) are illustrated.

The first phase is part of Figure 11.3; phase II illustrates that the grower has to choose between different varieties (or cultures) and agricultural processes. In this matrix all alternatives are evaluated according to the quality characteristics and the calculated importance, which are results of phase I. A result of II would be the identification of promising alternative varieties and agricultural production processes according to all the identified customer needs and wants. The roof of phase II illustrates that varieties and the agricultural production processes are not independent. The matrices may be defined for numerous purposes, e.g., the synergy between variety and process, or the synergy between varieties, e.g., season or pollination. In Figure 11.4, varieties and agricultural processes are illustrated side by side; alternatively, the agricultural processes could be evaluated given the certain varieties, but as research and development are often related to one of them, the separate or par-

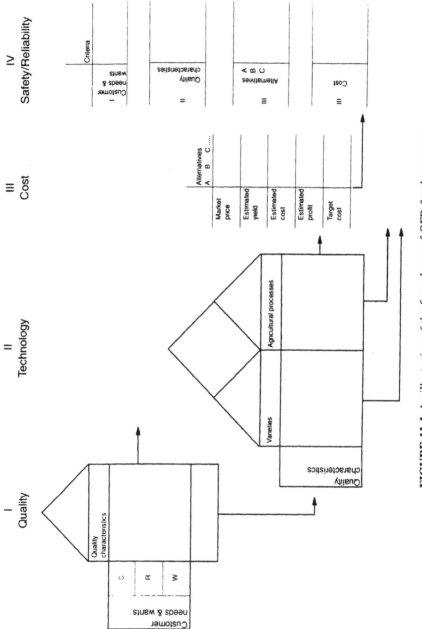

FIGURE 11.4 An illustration of the four phases of QFD for the grower.

215

allel approach has been chosen. If we only had to consider quality from a customer point of view, the quality function deployment would end after this phase; however, we do have to take both cost, safety and reliability into account at all levels in phase I and II. In this connection, phase III and IV are conditioned on phase I and II. Phase III and IV include all the dependent variables that are important to the grower and society. Cost and yield may be related to varieties and productions process (as rows below matrix II) and to the combination, in this case the alternatives A-C (in a new matrix). It is also possible to include target costs and then try to develop production processes that satisfy this criterion. Correspondingly, the safety and reliability deployment is the systematic assessment of all aspects of the QFD process on these criteria; HACCP (Hazard Analysis and Critical Control Points) principles may be used to identify the criteria. In the case of primary production, the reliability of the cost criteria could be included in the cost matrix as well. After completing the four phases, production and quality assurance planning follows.

In reality many decisions are conflicting. With QFD as a model for product development the choice of, e.g., the variety of a special culture would be conditioned on whether the variety possesses attributes that the consumer and the other customers appreciate, and whether the variety can be grown with a reasonable stable yield and profit for the grower. In the same way, new varieties and agricultural processes will be evaluated according to the defined quality characteristics. Both examples may be contrary to the traditional company-driven approach where primary economic and agricultural decision criteria would be used.

EXAMPLE—HOUSE OF QUALITY FOR STRATEGIC PEAS

This example (Bech et al., 1997a) is part of a large research project named "Strategic peas." The results are analyzed at both the measurement level and the latent level. As there is good agreement between the analyses, only the measurement level is reported here. With reference to Figure 11.3, this example is a part of the upper-right central relationship matrix where we have the consumer needs and perception of good sensory quality of deep-frozen green peas translated into sensory attributes, measurable by traditional sensory descriptive analysis (Hootman, 1992).

It is assumed that the translation of consumers' perception of the sensory quality can be done by the use of sensory descriptive analyses based on the perceptions of trained panelists. It is also assumed for both the panelists and the consumers that their perception is a function of three factors, which were selected in order to maximize the variation of the pea quality. The three factors, which form an experimental design, were size (4 levels), content of sucrose (2 levels) and surface color (2 levels, L-value from the HunterLab tristimulus L*a*b-measurements). After screening 42 batches of peas from the Danish 1993 harvest, 16 pea samples were selected according to this design. However, two of the 16 combinations were not available, so it was not possible to reach a completely balanced design. But still it is possible to estimate all the main effects and the two factor interactions considered most important.

Two consumer surveys were conducted in Denmark, a qualitative study in the form of four focus group interviews followed by a quantitative study including an in-home product test. The purpose of the focus groups was to gain information on consumer attitudes and behavior in relation to vegetables in general and specific knowledge of consumer needs and wants regarding deep frozen green peas, in QFD terminology *listen to the voice of the customer* (Griffin and Hauser, 1993). The results were used for development of the questionnaire for the quantitative study. Consumer needs and wants regarding good sensory quality of peas are listed in the left part of Figure 11.5.

The results from the consumer evaluations as well as the evaluations by the trained panelists were analyzed by using General Linear Models (GLM for Unbalanced ANOVA, SAS) with the evaluations as dependent variables and the experimental factors as the independent variables; the interactions considered most important were also modeled. For all significant effects the differences were analyzed in detail in order to explain the direction of the significant effects. The results from the consumer study and the sensory descriptive study are illustrated in Figure 11.5, the left part of the central relationship matrix and the upper part of the roof of the House of Quality, respectively.

By use of matrix algebra, the consumer requirements regarding good sensory quality are translated into sensory attributes measured by trained panelists. The first element of the right part of the central relationship matrix is a result of the sum of the multiplication of each element in the corresponding row with each element in the corresponding column (roof). In this case, the consumers' perception of the ideal smooth surface of peas is related to the size of the peas, where size 1 is preferred, and to the content of sucrose, where the high level (H) is preferred. The

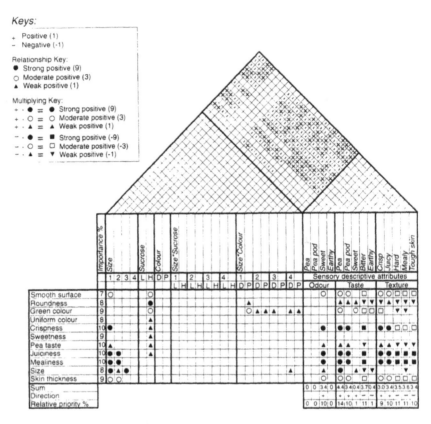

FIGURE 11.5 House of Quality for peas as part of a hot meal and with the involved consumer as target group. [Reprinted from *Food Quality and Preference* (Bech et al., 1997a), with pemission from Elsevier Science.]

first column in the roof illustrates the panelists' perception of pea odor as a function of the design factors. This column is empty, as the sensory attribute pea odor is not a valid discriminator between the different peas in the study. Consequently, there is no relationship between the consumers' perception of smooth surface and the panelists' perception of pea odor. This summing and multiplication continue until the final element down in the left corner (skin thickness/tough skin). At the bottom, we find a sum below each column. This is a weighted sum of the importance column and the actual column. As an example, the sum (3.4) below the third sensory descriptive attribute sweet odor is calculated as the following: $0.08 \cdot 3 + 0.07 \cdot 0 + 0.09 \cdot 0 + 0.08 \cdot 0 + 0.10 \cdot 9 +$

$0.09 \cdot 0 + 0.10 \cdot 1 + 0.10 \cdot 9 + 0.10 \cdot 9 + 0.08 \cdot 1 + 0.09 \cdot 3 = 3.39$. Finally, the relative priority of the sensory attributes is calculated.

The conclusion is that the consumers' perception of good sensory quality of green peas can be translated into sensory descriptive attributes. The consumers want peas with pea and pea pod taste, without bitter taste; the texture should not be hard and mealy or with tough skin, instead juicy and crisp texture is required; sweet odor is also a positive attribute. The two attributes sweet taste and earthy taste are in this case only of minor importance, while the other three odor attributes are of no value.

This research is an example of a general method for translating consumer wants regarding good sensory quality of food into sensory attributes measurable by trained panelist and with experimental design and ANOVA as intermediate methods.

CONCLUSIONS

House of Quality and Quality Function Deployment represent an integrated view of quality at the strategic, tactic and operational business level. QFD can and should be adapted to individual companies, teams and projects. The benefit of House of Quality is strengthened by the use of theoretical approaches to consumer behavior and food quality.

This chapter describes QFD and House of Quality with special attention to fruit and vegetables, where quality changes and customer needs and wants from all participants in the chain from consumer to producer have to be considered. This integrated House of Quality provides a framework for understanding the basis for decisions made so far and for understanding the potential of an integrated view of quality. The integrated view represents a complicated but realistic approach for marketing, production and development of fruit and vegetables with potential for success. It is not realistic to study all parts of the integrated House of Quality; alternatively, it can be used for the planning.

What kind of information do we have, where does it come from and from where do we need more information in order to develop fruit and vegetables that meet or exceed the consumers' and other customers' expectations? Until now research on consumer behavior and consumer needs has played a minor role in the research and development of fruit and vegetables. Much research has been done on postharvest and storage conditions as well as agricultural processes, these results and results of future research will have a much greater potential for successful implementation if a link is established to consumer research, the consumer needs and benefits.

Successful implementation of QFD requires commitment of the organization and it is most likely that the advantages will be strategic, in the form of long-term competitive advantages as a better understanding of customer needs and cross-functional team work, benefits that might be spread in the company (Griffin, 1992). Besides, the team's different professional skills will be of great importance. Research and attention to all the measurement areas in QFD are needed. The dependencies between measurements should not be overlooked. In order to study the latent independent factors, a quantitative approach is taken by Bech et al. (1997a). An interesting research area is also a formal modeling of the Kano model of must-be, proportional and exiting quality characteristics.

The development of fruit and vegetables using QFD should not be a substitute for the way it is done today. Rather, QFD should be seen as a supplement to existing methods and it is still important to utilize the technological and agricultural competencies. In this way, research and product development will become more goal-oriented, and there will be a more appropriate utilization of the involved competencies, resulting in products fulfilling or exceeding consumers' expectations to a greater extent.

REFERENCES

Akao, Y. 1990. *Quality Function Deployment Integrating Customer Requirements into Product Design,* Productivity Press, Portland, OR.

Bech, A. C., Engelund, E., Juhl, H. J., Kristensen, K., and Poulsen, C. S. 1994. QFood—Optimal design of food products. MAPP Working Paper No. 19, The Aarhus School of Business.

Bech, A. C., Hansen, M., and Wienberg, L. 1997a. Application of House of Quality in translation of consumer needs into sensory attributes measurable by descriptive sensory analysis. *Food Quality and Preference* 8: 329–348.

Bech, A. C., Kristensen, K., Juhl, H. J. and Poulsen, C. S. 1997b. Development of farmed smoked eel in accordance with consumer demands, in *Seafood from Producer to Consumer, Integrated Approach to Quality,* J. B. Luten, T. Børresen and J. Oehlenschläger, eds., Elsevier Science, New York, pp. 3–19.

Bech-Larsen, T., Nielsen, N. A., Grunert, K. G., and Sørensen, E. 1996. Means-end Chains for Low Involvement Food Products—A Study of Danish Consumers' Cognitions Regarding Different Applications of Vegetable Oil. MAPP Working Paper No. 41. The Aarhus School of Business.

Cardello, A. V. 1994. Consumer expectations and food acceptance, in *Measurements of Food Preferences*, H.J.H. MacFie and D.M.H. Thomsen, eds., Blackie Academic and Professional, London, pp. 253–297.

Charteris, W. 1993. Quality function deployment: a quality engineering technology for the food industry. *J. Society for Dairy Technol.*, 46(1): 12–21.

Cohen, L. 1995. *Quality Function Deployment How to Make QFD work for You*, Addison-Wesley, Reading, MA.

Dalen, G. A. 1996. Assuring eating quality of meat. *Meat Science*, 43(S): S21–S33.

Day, R. G. 1993. *Quality Function Deployment: Linking a Company with Its Customers*, ASQC Quality Press, Milwaukee, WI.

Day, G. S. and Wensley, R. 1988. Assessing advantage: A framework for diagnosing competitive superiority. *J. Marketing*, 52(2): 1–20.

Earle, M. D. 1997. Changes in the food product development process. *Trends in Food Sci. & Technol.*, 8: 19–24.

Feigenbaum, A. V. 1961. *Total Quality Control*, McGraw-Hill Book Company.

Gattorna, J. L. and Walters, D. W. 1996. *Managing the Supply Chain A Strategic Perspective*, Macmillan Business, Old Tappon, NJ.

Griffin, A. 1992. Evaluating QFD's use in US firms as a process for developing products. *J. Product Innovation Management*, 9: 171–187.

Griffin, A. and Hauser, J. R. 1993. The voice of the customer. *Marketing Science*, 12(1): 1–27.

Grunert, K. G., Baadsgaard, A., Hartvig Larsen, H., and Madsen, T. K. 1996. *Market Orientation in Food and Agriculture*. Kluwer Academic Publishers, USA.

Gustafsson, A. 1993. QFD and Conjoint Analysis. The Key to Customer Oriented Products. Thesis No 393, Linköbing Studies in Science and Technology, Linköbing.

Hauser, J. R. and Clausing, D. 1988. The House of Quality. *Harvard Business Review*, 66(2): 66–73.

Hofmeister, K. R. 1991. Quality function deployment: Market success through customer-driven products, in *Food Product Development from Concept to the Marketplace*, E. Graf and I. S. Saguy, eds., Chapman & Hall, New York, pp. 189–210.

Hootman, R. C. 1992. Manual on Descriptive Analysis Testing for Sensory Evaluation. ASTM manual series: MNL 13, Philadelphia.

Mazur, G. 1987. *Attractive Quality and Must-be Quality*. Goal/QPC. Translation of Kano, Seraku, Takahashi, Tsuji 1982 from Japanese.

Mazur, G. 1994. Quality Function Deployment—An Overview. Quality Function Deployment—For the Food Processing Industry. MI.

Mizuno, S. and Akao, Y. 1994. *QFD—The Customer-Driven Approach to Quality Planning and Deployment*. Translation from original in Japanese (1978), Asian Productivity, Tokyo.

Oliver, R. L. 1997. *Satisfaction. A Behavioral Perspective on the Consumer*. The McGraw-Hill Companies, Inc., New York.

Olson, J. C. and Jacoby, J. 1972. Cue utilization in the quality perception process, in *Proceedings of the Third Annual Conference of the Association of Consumer Research*, M. Venkatesan, Ed., Association for Consumer Research, Chicago, pp. 167–179.

Pedi, R. and Mosta, R. 1993. Total quality in product development. *Preferred Foods*, 162(2): 44–46.

Poulsen, C. S., Juhl, H. J., Kristensen, K., Bech, A. C., and Engelund, E. 1996. Quality guidance and quality formation. *Food Quality and Preference*, 7(2): 127–135.

Shewfelt, R. L., Erickson, M. C., Hung, Y.-C., and Malundo, T. M. M. 1997. Applying quality concepts in frozen food development. *Food Technol.*, 51(2): 56–59.

Shewhart, W. A. 1931. *Economic Control of Quality of Manufactured Products*, Van Nostrand Company, New York.

Skytte, H., Grunert, K. G., Jensen, N. B., Mortensen, J. B., and Brunsø, K. 1997. Unpublished results. The MAPP Centre, The Aarhus School of Business, Denmark.

Sloof, M., Tijskens, L. M. M., and Wilkinson, E. C. 1996. Concepts for modelling the quality of perishable products. *Trends in Food Sci. & Technol.*, 7: 165–171.

Steenkamp, J.-B. 1989. *Product Quality: An Investigation into the Concept and How It Is Perceived by Consumers*,. Van Gorcum, Assen/Maastricht.

Surak, J. G. and McAnelly, J. K. 1992. Educational programs in quality for the food processing industry. *Food Technol.*, 46(6): 80, 83–85,87,89–90.

Swackhamer, R. 1995. Responding to customer requirements for improved frying system performance. *Food Technol.*, 49(4): 151–152.

Sørensen, E., Grunert, K. G., and Nielsen, N. A. 1996. The Impact of Product Experience, Product Involvement and Verbal Processing Style on Consumers' Cognitive Structures with Regard to Fresh Fish. MAPP Working Paper No. 42. The Aarhus School of Business.

Urban, G. L. and Hauser, J. R. 1993. *Design and Marketing of New Products*, 2nd ed., Prentice Hall International Editions, Paramus, NJ.

COMMON GROUND

- Consumer acceptability is more difficult to measure than fruit and vegetable quality characteristics.
- The lack of meaningful measures of consumer acceptability limits our ability to improve fruit and vegetable quality.
- The lack of common terminology limits integrated approaches to quality improvement.

DIVERGENCE

- Definition and integration of quality and acceptability.
- The ability to adequately measure consumer acceptability that will provide meaningful information to breeders and handlers.
- The importance of economics and postharvest physiology in understanding quality and acceptability of fresh fruits and vegetables.

FUTURE DEVELOPMENTS

- Commonly accepted terms of quality and acceptability.
- Quantitative measures for consumer acceptability.
- More integrated studies to improve fruit and vegetable quality.

AN INTEGRATED VIEW

CONTEXT

- The marketplace is complex and problems cross disciplinary lines and perspectives.
- Academic research tends to be disciplinary in nature.
- Terminology is a major impediment to truly integrated approaches.

OBJECTIVES

- To emphasize the importance of economic considerations in quality assessment of fresh fruits and vegetables and present approaches for integrating economics and quality.
- To present the IQM model, which integrates quality assessment of processed vegetables with consumer requirements.
- To incorporate use of soft-systems methodology as a basis for integrating research on fresh fruit and vegetable quality.
- To use a problem-based approach as a point of integration for research on fruit and vegetable quality.
- To provide a set of terms and concepts that will help serve as a basis for integrated approaches to fruit and vegetable quality.

CHAPTER 12

Economics of Quality

WOJCIECH J. FLORKOWSKI

A CONSENSUS about what is quality seems elusive. The lack of a universal definition signals the complexity of the issue and the variety of perspectives that exist. Because quality evolves with changing technology, markets, industry structure, and consumer preferences, a single, all-encompassing definition applicable over an extended period can incorporate only common elements.

Hill (1996) offers a broad definition that meet these criteria, stating that quality is a composite attribute of the products that have economic or aesthetic value to the user. This definition represents a consensus of task force members deliberating quality of 12 groups of agricultural products. Markets reward quality and instantly transmit market signals between buyers and producers. Shewfelt et al. (1997) review definitions of quality for a specific group of users—retail consumers. Their definition reflects the importance of consumer expectations and the anticipation of changing preferences.

The broad quality definition offered by Hill and the narrow definition developed by Shewfelt reflect the ability of markets to measure quality, rewarding the best and discounting the worst. The interpretation of quality has consequences for empirical research in economics, market-

Paper prepared for presentation at the International Conference "An Integrated View of Fruit and Vegetable Quality," Potsdam, Brandenburg/Germany, May 10–May 15, 1997.

ing, and management. The complexity of measuring quality requires the development of a synthetic market description that is easily applicable and clearly understood by participants. With the development of measuring technology, we can account for minute changes in specific external and intrinsic attributes. Information from the attribute measurement does not represent equal value to the market. Markets select and rank attributes according to supply and demand conditions. The ability to measure an attribute is unimportant, despite its presence and technical feasibility, if buyers do not recognize such attributes as contributing toward the value of the product. In this chapter I review the dynamic nature of fresh produce quality and its interpretation by various links in the marketing chain. Recognizing the need for a multidisciplinary approach to quality, I stress the economic importance of the quality to fresh produce industry. Constant industry demand for practical solutions and recommendations is generated from empirical studies applying economic theory and statistical methods modeling purchasing and consumption behavior. Each of the described approaches has its advantages and limitations, but can provide unique insights about the market and consumers to benefit research in other disciplines, public policy, investment, and management decisions.

QUALITY DYNAMICS

The differentiation of agricultural and food products according to various quality attributes has been a recognized trade practice since the inception of exchange. In response to market demand farmers supply produce with attributes wanted by the ultimate consumer. Manufacturing requirements of the processing industry pressure farmers to deliver commodities with quality attributes that maximize processors' profits. Commodities with preferred attributes make the processing efficient and yield a consistently uniform product. Bulk agricultural commodities are particularly suited for industrial processing. Industrial food processors purchase grains, oil seeds, and later, livestock and milk from farmers according to their physical and chemical attributes. The presence and the intensity of an attribute lead to classification of the supplied product and the adoption of quality-oriented pricing. For example, attributes subject to inspection against manufacturing requirements applied to vegetable crops include quality-specific measurements of potatoes such as starch content.

Quality of vegetables and fruits destined for processing differs from the fresh produce sold at retail outlets. The type of marketing channel selected by growers before planting influences farms' resource allocation. The marketing chain for fresh produce is shorter than for processed produce reflecting the perishability. Farms near major urban centers can sell fresh produce through traditional channels or invite customers to "pick-your-own" (PYO) enterprises thereby minimizing the length of the marketing channel. PYO farms may revise the importance of quality attributes. Results of a recently conducted survey of peach growers show that commercial growers and pick-your-own operators emphasize different quality attributes of peaches (Florkowski et al., 1997). Commercial growers-shippers stress size and firmness. Standards for grades used in wholesale trade specify size as an attribute influencing classification, while firmness facilitates shipping. Pick-your-own operators considered color and maturity (the absence of firmness) as the primary quality attribute. Clearly, both types of growers were responding to quality requirements expressed to them by buyers. If the signals were correctly transmitted, then quality attributes expected by consumers vary depending on place of purchase, e.g., a supermarket vs. PYO farms. Poor signal transmission can cause sluggish retail sales because consumers cannot find peaches of expected quality. The examination of institutional arrangements along the marketing chain can provide additional explanation and lead to effective remedies.

GRADES AND STANDARDS

Grades and standards improve marketing efficiency by reducing the risk of miscommunication between sellers and buyers. Transaction costs also decrease because the physical inspection of each shipment becomes unnecessary. Wide application of standards for grades of homogeneous products contrasts with the specialized nature of standards for quality and grades in the fresh produce industry. Grades require uniformity of characteristics such as size, color, or cultivar. Stone fruit or citrus fruit, berries or melons, root vegetables or leafy vegetables may differ in their attribute mix to the extent that a standard or a grade may only broadly describe requirements and the condition required for fresh produce may be product specific.

In the United States, the use of standards for grades is generally voluntary in fresh produce marketing. The initiative in the development of grades or standards could determine the extent of control over the ultimate set of standards applied in marketing channels. If quality characteristics included in the description reflect only the view of the initiating party, some economically important attributes may be ignored. Deliberate or unintentional omissions may hurt the ability to effectively communicate among various links in the marketing chain rewarding attributes to which retail consumers attach little importance and are unwilling to pay a premium.

How well standards reflect the quality attributes desired by users and recommended by market is often a cause of debate. Marketing channel participants allege sometimes that another party takes an unfair advantage of the process. Government investigations and possible interventions could be justified if the economic loss caused by confirmed unfair behavior exceeds the cost of implementing a change, therefore increasing the overall welfare.

Enforcement of Standards for Quality

The enforcement is largely dependent on the roles played by the private sector and the government in recognizing the need for and benefits of standards for quality grades. The voluntary use of standards for grades transfers the responsibility for enforcement to buyers and sellers, away from the government. Specification of a grade and naming the source of its description in a contract secures rights to legal recourse should the quality of delivered produce be questioned. The ultimate enforcer of quality standards on the fresh produce market is the retail customer.

Leaving the enforcement of standards for quality to market forces is cost-effective. However, not all attributes essential to retail customers, and therefore to all links in marketing chain, can be evaluated at the point of purchase through visual inspection. Nichols (1996) noticed that intrinsic attributes are not reflected in grading systems and are excluded from fresh produce standards. This gap leaves a place for government as a monitoring, regulatory, or even enforcing agency.

An issue receiving particular attention and subject to government regulations is the testing for pesticide residue and monitoring of pesticide use in production. Pesticide use and residue regulations are intended to protect consumers from unsafe food. Food safety is of primary concern to society and represents a condition that must be met before other quality attributes can be considered.

In recent years, the emphasis on produce quality has increased in the American supermarket. Produce departments generate a substantial portion of the overall revenue. Therefore, the supermarket name becomes synonymous with produce quality and quality becomes an instrument of marketing strategies aimed at consumers. A recent study (Blisard et al., 1999) showed that expenditure on fresh fruit and vegetables in the total household food expenditures increases as incomes rise. Fresh produce quality enables growers and marketers to overcome barriers to individuals' consumption growth and slow population growth, while satisfying increasingly sophisticated consumers. The picture painted here applies to high-income economies because the primary driving force behind changing product and quality selection is the increasing income.

INTERPRETATION OF QUALITY

The demand for quality attributes encouraged the investment in research addressing quality problems encountered in production and marketing channels that limit sales, profits, consumption, and consumer satisfaction. The identification of economically important quality issues is based on the highest expected returns. The priority setting determined by scientists must be reconciled with an industry viewpoint, which is driven by the necessity to remain profitable. Industry needs include empirical studies of demand and markets to provide an overview of a fresh produce sector for the purpose of strategic decision making. Information generated by economic studies influences resource allocation and investment decisions. Consumer studies provide unique insights about behavior and facilitate management selection of marketing techniques. Further, consumer education may enhance the understanding of changes in the production, handling, and distribution by communicating benefits of new solutions.

The complexity of improving produce quality attributes encourages a multidisciplinary approach to quality. The multidisciplinary approach, however, conflicts with the reward system based on individual and disciplinary effort. The existing reward system encourages fragmentation of research effort and leads to delays in sharing important discoveries because of the desire to claim the disciplinary recognition of the discovery through publications in peer-refereed outlets. The fuzzy delineation of individual contributions in addressing produce quality requires frequent adjustments to ensure a balanced distribution of rewards or face potential waste of resources on duplication of efforts. Duplication cannot always be avoided, while replication of research discoveries is necessary to commercialize results. Society generally accepts potential loss of resources spent on similar projects although attitudes change, and financing research from public funds becomes an issue in times of fiscal prudence.

Fresh produce quality is frequently addressed by individual disciplines because of the highly specialized nature of a required solution. The question that lingers is, How are research results disseminated to other disciplines and what are the incentives to learn about each other's recommendations? Practitioners may face several options to address a problem, but the disconnected individual solutions lead to inefficiencies, which are discovered only through trial and error.

MARKET VALUATION OF QUALITY

The neoclassical framework of pricing quality attributes implies that the final price reflects prices of the bundle of attributes contained in a unit of a product. Markets are assumed to be perfectly competitive with unimpeded flow of information instantly available to all market participants. Consequently, prices change instantly in reaction to incoming information. Many agricultural commodity markets were modeled as perfectly competitive, even though some required adjustments to account for biological production lags and inventories.

In case of fresh produce, the bundle of attributes is represented by a single fruit or a vegetable, or a trade prescribed unit. Fresh produce marketing takes place at various market levels, which redefine the basic unit. For example, a farmer may deliver a truckload of tomatoes to a packer, which are packaged in boxes for shipment to a wholesaler, but are sold as individual items at the retail outlet. Under the neoclassical framework, perfect transmission of market signals is necessary for multilevel market analysis because it treats the wholesale and the farm level demand as the demand derived from consumer demand demonstrated at the retail level.

Quality interpretation and measurement based on the needs of various agents throughout the marketing chain prevents the application of a single theoretical approach to price attributes. The assumption of perfect competition implicit in many economic analyses of price-quality relationship represents an approximation of reality. Predicting behavior is essential for practical applications and, despite its limitations, economic analysis provides a benchmark for making informed decisions. Statistical testing verifies the explanatory power of economic models and relates the incidence of outcomes to preselected probability levels. Practitioners have to reevaluate outcomes in response to the steady flow of information not captured by a model using their market knowledge and experience.

Hedonic Pricing

In the agricultural economics literature, the vast majority of studies applied Hedonic technique to pricing of grains, oil seeds, and fiber, or to meat and dairy products. Although Waugh (1928) pioneered the application of Hedonic technique to price quality attributes of fresh vegetables decades ago, the technique has been used infrequently. The Hedonic technique studies consumer demand for attributes by applying the utility maximization framework as the foundation of economic analysis. Much of the theoretical work on consumer utility maximization was conducted by Theil (1952), Lancaster (1966), and others, and provided sound theoretical foundations for empirical studies.

Consumers view a good as a bundle of attributes and rationally evaluate how each of them contributes to the level of satisfaction or utility. An attribute contribution is measured "on the margin," i.e., by how much the utility level changes from consumption of an additional unit of an attribute. An algebraic example of the application of the Hedonic technique consists of the utility maximizing equation:

$$\underset{\max}{U} = f(X_i; K_j) \qquad (12.1)$$

where X_i is the vector of produce attributes and K_j is a set of other relevant variables influencing consumer utility associated with produce consumption. Empirical studies include in the K set of variables demographic and socioeconomic measures, descriptors reflecting preferences, and constructs based on beliefs or attitudes of consumers. The differentiation of Equation (12.1) yields a first-order condition which, when solved for the product price, provides a measure by how much a price changes in response to the varying amount of an attribute in a traded unit of a product. The marginal quantity of an attribute is necessary to calculate the marginal price per unit and the calculation of the marginal monetary value of the total amount of the attribute contained in the product. Hedonic technique implies that the price paid by buyers is a sum of the marginal monetary values of all attributes.

Consumers recognize that the mix of attributes varies across units of fresh produce. The tradeoffs between one attribute and another may influence consumers' choice. The Hedonic technique and the utility maximization framework permit the calculation of the marginal rate of substitution between expenditures and the volume of an attribute. Therefore, results of Hedonic analysis provide insights about the importance of specific attributes to buyers expressed in terms of money, a measure easily understood by all parties involved in marketing fresh produce.

Economists use the Hedonic technique at various market levels, e.g., farm, wholesale, or retail. Farm- or wholesale-level demand for attributes is assumed to be derived from the retail level. This assumption holds when the marketing margin remains unchanged for all prices and quantities of the product in question. It implies that slopes of farm, wholesale, and retail demand curves are identical. In reality, some variation occurs, but it is generally recognized that the demand curves have similar slopes over the studied range of prices and quantities. Violation of this assumption renders the analysis invalid and of little practical relevance.

Various studies of Hedonic techniques have been based on time series and cross-sectional data. Jordan et al. (1985) used data collected on tomato prices registered on a single day assuming the fixed supply of produce and its characteristics. The use of cross-sectional data to price attributes of fresh produce closely reflects market conditions. The assumption of fixed supply is plausible for studying consumer demand for produce with short shelf life. The condition of leafy vegetables or root vegetables sold with leaves deteriorates particularly fast because of a large transpiration surface. Other vegetables, including many root vegetables with other plant parts detached, may remain in good condition for an extended period of time if properly handled.

Rosen (1974) applied Hedonic technique to time series data, where product supply responds to the demand for a quality attribute over a period of time. Supply is no longer fixed and attribute prices can be estimated by a system of supply-and-demand equations. Data requirements for an empirical application of this approach are considerable and frequently insufficient to conduct an empirical study. Results based on a simultaneous system of supply-and-demand equations estimated using time series data provide insights about the structural nature of the market, but may be inadequate to explain behavior observed on any particular day. On the other hand, cross-sectional data capture the situation in a particular moment in time, which is assumed to be continuously replicated. Users of information stemming from Hedonic price studies should be aware of limitations of each approach. Weather, cultivar replacement, seasonality, multiple grades and standards, poor record keeping of produce moving along the marketing chain, and other factors disrupt observations and limit the available data.

The functional form of the estimated Hedonic equations is determined empirically because theory does not provide selection guidelines. Past studies referred to various methods of selecting the preferred functional form. For example, Jordan et al. (1985) used transformation proposed by Box and Cox (1964) and applied by Huang (1979) to the estimation of the demand function, because this approach offered flexibility for testing various functional forms. Brown and Ethridge (1995) noted that linear functional forms fitted better aggregated data (these are often national time series statistics), whereas the nonlinear form suited disaggregated data. Wahl et al. (1995) used the P-test to test data compatibility of five functional forms of the Hedonic price model. The functional forms included linear, semi-log, double log, inverse log, and quadratic specifications. Besides the use of statistical techniques, visual assessment of plotted data can provide hints about the preferred functional form of a Hedonic equation.

The primary result of empirical application of Hedonic technique is determining the role product quality attributes play in explaining the price and not in the ability to predict buyers' future behavior. Furthermore, because a product represents a bundle of attributes and fresh produce is versatile in consumption, retail consumers may price the same attribute differently given the intended use. Consumer preference dynamics make the analysis at the retail level more difficult than at the wholesale or farm level where buyers' preferences seem to change with less frequency. In an attempt to improve the validity of empirical analysis, economists choose to work with large number of observations to compensate for variability in utility derived from purchase of a good.

Contingent Valuation (CV)

CV technique has been applied to nonmarket good valuation and benefits derived from noncommercial objects. This technique is used in environmental studies and is an extension of cost-benefit analysis of alternative policies aimed at protection of natural resources. In recent years, formulation of environmental policies and search for practical solutions have given a boost to CV studies.

CV assumes that resources have use and nonuse value. The use value occurs during the actual use (Kriström, 1990), and the nonuse value is attached to a good by consumers. Fresh produce quality attributes include the use and nonuse value justifying empirical CV studies. Consumer preference for fruit color is an illustration of enjoyment derived from an external attribute, which may be unrelated to other physical characteristics or intrinsic attributes relevant to eating. The past use of alar in American apple orchards to enhance fruit color is an example of growers' response to consumer color preferences rather than an attempt to influence eating quality. Misra et al. (1991) showed that consumers preferred pesticide-free produce, but had little tolerance for visible defects. Whereas visible quality defects are noticed by consumers, applications of pesticides are perceived to be safe and necessary to produce crop.

Empirical applications of the CV technique are based on the neoclassical theory of utility maximizing consumer. Although the theoretical framework is similar to that applied by Hedonic pricing technique, the lack of actual market prices reflecting the demand for nontraded goods places a particular burden on the researcher to clearly define the relationship between the price of an attribute of a nontraded commodity and explanatory variables. The CV approach assumes that consumers' choices reflecting utility, or satisfaction derived from the consumption of produce, are bounded only by exogenously determined income and the market prices of all goods. The marginal utility of each quality attribute is derived from the utility maximizing equation and the income constraint. Moreover, the marginal utility obtained from consumption of each additional unit of attribute is positive, but decreasing. For example, within the same meal each eaten strawberry satisfies the consumer, but the satisfaction from eating the first strawberry is higher than from all berries eaten afterwards.

The concept of utility imposes strong limitations on the economic analysis and recommendations from empirical CV investigations. Theoretically, consumers are assumed to be able to recognize the choice maximizing their utility, to order all choices according to the level of utility they provide, and to assess the additional amount of utility gained or lost by choosing among available alternatives. The abstract nature of utility permits only its indirect measurement. Typically, empirical studies assume that the decision to purchase or to abstain from a purchase mirrors the utility derived from the product. The decision is irreversible and maximizes the satisfaction of a rational consumer.

The CV technique is useful in researching the quality of fresh produce and buyers' behavior because it allows quantification of attribute effects on purchase decision probability before a product with a specific mix of characteristics becomes available to consumers. Many CV studies focus on how much consumers would be willing to pay above the going market price for certain attributes. Some argue that empirical studies should distinguish between acceptability and willingness-to-pay. Organic produce or fresh produce grown under Integrated Pest Management (IPM) regime are examples of goods still little known to consumers, while various opinions circulate about crops with the attribute mix altered through genetic engineering. An acceptability study may be a prerequisite for willingness-to-pay investigation.

Consumer purchase or consumption decisions can be described by a binary variable. The variable assumes the value of one if the decision to purchase was made and equals zero otherwise. The categorical nature of the dependent variable presented a problem in the application of statistical methods and the interpretation of estimation results. The theoretical framework and the appropriate estimation technique were reconciled by McFadden (1981). McFadden's approach transforms the algebraic expression of utility into a statistical equation that can be estimated by probit or logit technique. Probit technique is considered superior to the logit approach because its application requires rigorous distribution assumptions. The logit approach relaxes the normality assumption and, by rule of thumb, empirical logit models provide similar results if estimated using large data sets.

Factors that influence the decision to purchase may be different from those responsible for a premium a consumer is willing to add to the comparable price. Consumer surveys probing for willingness-to-pay ask respondents to choose from several price premium alternatives. The selection of a premium begins at the value of zero and is constrained by investigators at a level perceived as the reasonably high. The selection of each alternative is coded as categorical variables assuming values of zero to record the choice of the first alternative, two to record the choice of the second alternative, and so on. This coding scheme requires a different estimation technique if a categorical variable becomes the dependent variable. Ordered probit or ordered logit techniques are applied and estimated coefficients interpreted as indicators of a probability of the respondent's choice falling into a particular category describing the range of premiums.

In the case of all four techniques (i.e., probit, logit, ordered probit, and ordered logit), the practical interest lies in the ability to identify the effect of each explanatory variable on the dependent variable. The marginal effects of quality attributes measure by how much the probability of paying a particular premium changes in response to the presence of an attribute. Marginal effects in ordered technique applications provide intuitively less obvious practical results than those obtained from regression estimations using a continuous dependent variable.

The lack of rigorous statistical tests and measurement inaccuracies associated with abstract concepts (e.g., utility, aesthetic attributes) allows dispute about the interpretation of results. The ability to predict consumer choice is limited to a decision to purchase in an isolated situation. Consequently, empirical studies turn to experimental economics in search of improving the predictive power of theoretical models. Experimental economics involves virtual situations and includes real incentives (in form of money payment) to elicit probable responses from subjects. The importance of recommendations based on economic experiments depends on the accuracy of the virtual environment in which subjects make their choices and their understanding or involvement in the problem. Because participants receive a payment as a reward, such studies can be expensive.

Attitudes, Quality Perceptions, and Behavior

Positive consumer attitudes are a prerequisite to sustained market demand for fresh fruit and vegetables. These positive attitudes are formed over an extended period of time and reinforced by numerous factors. Social norms, formally acquired knowledge, or experience may dictate attitudes, but the variety of fresh produce on the stand, prices, convenience, and the expected preparation time may ultimately lead to choices that are logical from the standpoint of a consumer but inconsistent with the declared preferences. The complexity of predicting behavior at a retail produce purchase is the context and the timing of the observed selection.

In industrialized countries the attitude toward produce consumption has been influenced by new discoveries in nutrition and medicine. However, large variability in perceptions persists. For example, 58% of consumers surveyed in Berlin in 1994 expressed the opinion that they would like to eat more fresh vegetables than they actually consumed (Brückner et al., 1996) as compared to 34% of residents of the metropolitan Atlanta area (Lai et al., 1997) (Table 12.1). In another study, about 80% of surveyed Japanese consumers felt they should eat more vegetables, while only 20% of British respondents expressed a similar view (Moteki and Muller, 1992). It appears that in these cases, positive attitudes to-

Table 12.1 Farm Value of Selected Vegetables and the Reported
Purchase Frequency by Consumers in Atlanta and Berlin.

Rank	Vegetable	U.S. Farm Value (mln $, 1989)	Rank Based on Percent Purchasing Often or Very Often	
			Atlanta	Berlin
1	Tomatoes	1,824	1	2
2	Lettuce	950	8	4
3	Onions	538	2	1
4	Sweet corn	468	—	6
5	Carrots	297	3	3
6	Broccoli	276	9	5
7	Snap beans	228	5	7
8	Cauliflower	205	6	11
9	Cucumbers	203	4	9
10	Leeks		7	—
	Brussels sprouts		10	13
	Asparagus		11	12
	Green peas		12	10
	Mushrooms		13	8

Sources For Atlanta—Dept of Agricultural and Applied Economics, The University of Georgia, College of Agricultural and Environmental Sciences, Griffin Campus, Griffin, U S For Berlin—Institut fur Gemüse und Zierpflanzenbau, Grossbeeren, Brandenburg, The Federal Republic of Germany

ward fresh vegetables are not reflected in behavior and consumers purchased less vegetables than desired. In these examples, the product is defined as an aggregate "vegetables" and inferences based on the reported perceived adequacy of eating vegetables may be poor predictors of purchases of specific kind of vegetables. However, the general information may still be used in choosing marketing strategies and short-term management decisions such as implementation of generic promotional campaigns stressing the accessibility of fresh vegetables.

The importance of fresh produce purchases seems to be increasing because of the steadily intensifying messages stressing the positive role of produce to maintain human health. These messages may shape preferences and change attitudes by increasing consumer emotional involvement in the selection, purchase, and consumption frequency of fresh produce. Lack of involvement in the purchase decision tends to weaken the predictive ability of econometric models and severely impairs the practical significance of empirical research recommendations. The involvement in the purchase of fruit and vegetables seems to fluctuate in response to many economic and environmental factors and the availability of fresh, frozen, canned, or dehydrated substitutes.

Rational consumers demonstrate behavior by making a purchase. A "rational" consumer's selection reflects inner feeling or his or her attitude toward an object. Learning about consumer attitudes to predict purchase implies consistency between the nature of beliefs about a product and actually observed behavior (Ajzen and Fishbein, 1977). Knowledge of attitudes is useful in making decisions regarding an expansion of production, entry into a new market, or a change of a technology to supply produce with new attributes. The mutliattribute attitude model assumes that behavior is consistent with attitudes.

Ajzen and Fishbein (1977) noted a lack of theoretical basis linking attitudes and behavior. Consequently, investigators' intuition guides empirical research, which may be of limited theoretical value. Practitioners may gain some insights, but a study does not advance economic theory. Although the multiattribute attitude model was applied in numerous studies, there is a paucity of studies applying this approach to food products and fresh produce in particular. This gap in the literature can be attributed to the lack of data caused by difficulties in measuring attitudes and behavior. The need for highly disaggregated data makes them costly to collect.

Consumers treat a product as a combination of attributes, but believe that some attributes are critical. These salient beliefs determine attitudes. The strength of beliefs is critical in the multiattribute attitude modeling. Algebraically, the attitude model is:

$$\text{Attitude} = \sum_{i=1}^{n} \beta_i Z_i \ i = 1, 2, \ldots, n \qquad (12.2)$$

where the β's are weights attached to salient beliefs, and the strength of a belief that a product has an attribute i is represented by z's. The model treats all attributes as equally important, assuming that a consumer makes a proper assessment when asked to evaluate the product.

Because attitudes are inner feelings, they are measured implicitly using various scales containing several possible answers. A consumer indicates which of the offered options best reflects his or her attitude. The selection of an option becomes less obvious over time because other environmental variables may continually shape attitudes and their intensity. For example, freshness was among the most desired produce attributes reported by both Atlanta, U.S., and Berlin, Germany, consumers (Brückner et al., 1996; Lai et al., 1997), but freshness was found to be an expected attribute of many other foods and food products. While it is reasonable to assume that freshness will continue to be an important characteristic, other attributes such as size or uniformity may be less important to the whole population of consumers.

Multiattribute attitude model estimation depends on consumer memory and ability to recall experience with a product or any other information that could assist in the process. Because research shows that responses can be easily influenced by environmental factors and the thinking process, measuring attitudes remains difficult. Monitoring consumer environment can identify changes and strength of attitudes, but requires scarce resources.

Brand Names

Brand names are used successfully to establish a market share by convincing buyers that the branded product offers a level of satisfaction that cannot be provided by generic products. Through careful promotion and image building producers or suppliers of branded produce increase consumer loyalty and involvement in making the purchase decision. Because consumer involvement responds to various factors, some companies have established more than a single brand name in order to segment the market and maximize revenues. Branded fresh produce builds its market reputation by ensuring consistent quality by fully integrating production and marketing systems (Nichols, 1996). Product uniformity is ensured by the use of identical varieties and standardized cultural practices.

Tropical fruits including banana, pineapple, and citrus have been branded for years. Other food products with worldwide brands include mushrooms, spices, and beverages such as coffee, tea, and cocoa. Recent efforts to establish new banana brands proved that the task is costly and the brand may not be readily recognized or accepted by consumers. Erratic presence on the market is a sufficient factor to limit the effectiveness of promotion or advertising of brand-specific quality attributes and the association between the brand name and a product in the minds of consumers.

An example of supplying a branded fresh fruit is a new cultivar of pineapple developed using tissue culture. The fruit, marketed under a recognized brand name, has been promoted as being sweeter than other available cultivars. The company modified the original brand name used on pineapples of other cultivars sold by the company to reflect and promote its primary attribute—sweetness. The fruit is smaller in size than other pineapples and has an intense yellow color. The size of a pineapple can encourage purchases because a smaller size may appeal to small households. The yellow skin of the new cultivar can visually communicate to consumers the maturity, implying taste.

Among temperate fruits, apples are probably branded most often. In the United States apples are branded by growers' associations rather than a single producer. Other branded temperate-zone fruits and vegetables include pears, melons, onions, and carrots. A separate group of branded produce is represented by edible nuts and berries. Both in-shell and shelled nuts are sold under brand names, although the frequency of brand use varies among different nut industries. Generally, the brand name is used more often on nuts where the industry's internal organization led to highly integrated production and marketing, increasing the efficiency of brand promotion.

Successful branding emphasizes qualitative attributes (e.g., taste of a new pineapple cultivar) or nutritive characteristics (Nichols, 1996). A study of Berlin, Germany, consumers indicated that attitudes toward branding fresh produce are influenced by consumer characteristics and perceptions. Furthermore, the nutritive characteristics of branded produce, especially when reflecting different production methods, attract different groups of consumers, who can be distinguished from the general population. Organic produce and produce with scientifically proven nutritional or medicinal benefits will become increasingly popular among consumers.

QUALITY AS A RISK MANAGEMENT TOOL

Quality is a criterion distinguishing among substitutes and has been used effectively to establish market segments on durable goods markets. For the fresh produce industry, quality is important although the value of a single purchase of produce can be small and the risk of buying poor-quality product is perceived as low. However, Shewfelt's concern about quality defined from the standpoint of a retail customer implies the need to minimize quality defects because defects can induce a shift of demand away from a specific kind of fresh produce.

Market observations confirm that repeated negative experiences with the quality of purchased goods affect consumer purchase decisions or the volume purchased. The distinction between the purchase decision and the consumption frequency or volume consumed has been modeled using a double-hurdle model. The premise of the model's theoretical framework is that consumers consciously choose to participate in the market by making a purchase and, subsequently, make a separate decision about how much of a good to purchase. These two separate decisions are assumed to be determined by different sets of factors, some of which are the good's quality attributes. For example, Park and Florkowski (1999) showed that positive experiences with shelled pecans influenced the consumption frequency, while negative experiences lowered the probability of making the purchase decision. Furthermore, they found that consumers may be divided into two uneven-size groups, of those who regularly purchase the product and those who do not or do so only occasionally.

Variable reaction of consumers regarding their experience with product quality suggests the need to maintain the integrity of fresh produce or risk sales. A high-priced item requiring a substantial consumer involvement in the purchase decision may be promptly ignored by buyers if it does not meet quality expectations. Adverse selection behavior aimed at risk reduction can encourage purchases and consumption of standard quality fruit and vegetables sold at low prices. Furthermore, the promotional message aimed at one group of consumers may be unsuitable for other groups. To attract new buyers, rather than only maintaining the current population segment, the produce industry may reconsider the marketing strategy in the context of expected revenue changes.

Strategies addressing market risk may vary with the size of a farm operation. Demand for special or exotic produce can be met by small growers exploiting cross-cultural eating differences, which create market niches. Specialized produce often sells at higher prices and offers higher margins than sales of the most frequently eaten fruits or vegetables. By choosing the production of novelty produce in response to consumer demand, farmers may reduce market risk and isolate their niche from direct competition. But a sustained effort to ensure high and consistent quality produce is necessary to fend off the entry of competitors or to maintain a segment of buyers.

Price received by farmers for fresh produce represents only a portion of the marketing bill. Growing pressure to generate income in a market saturated with competing products leads to the use of produce quality attributes as a source of differentiation and identity. Flexibility in responding to market signals and the anticipation of new food and eating trends is necessary for sustained competitiveness within the fresh produce industry. The varying consumption frequency and volume consumed of specific fruits and vegetables requires sustained research of consumer preferences and the ability of markets to supply produce of expected quality.

REFERENCES

Ajzen, I. and Fishbein, M. 1977. Attitude-behavior relations: A theoretical analysis and review of empirical research, *Psychological Bulletin,* 84: 888–918.

Blisard, N., Smallwood, D., and Lutz, S. 1999. Food Cost Indexes for Low-Income Households and the General Population, USDA-ERS, Technical Bulletin No. 1872, 26 p.

Box, G. E. P. and Cox, D. R. 1964. Analysis of transformations. *Journal of Royal Statistical Society, Series B,* 26: 211–243.

Brown, J. E. and Ethridge, D. E. 1995. Functional form model specification: An application to hedonic pricing. *Agriculture and Resource Economics Review,* 24: 166–173.

Brückner, B., Schonhof, I., Florkowski, W. J., and Kuchenbuch, R. 1996. Qualitätserwartungen an Gemüse in Berlin, DGG, Erfurt, 28.2.-1.3. *1996 BDGL Schriftenreihe,* 14: 22.

Florkowski, W. J., Purcell, J. C., and Hubbard, E. E. 1992. Importance for the U.S. pecan industry of communicating about quality. *HortScience,* 27(5): 462–464.

Florkowski, W. J., Huang, C. L., Brückner, B., and Schonhof, I. 1994. Unpublished Survey Results of Atlanta Metropolitan Area Consumers.

Florkowski, W. J., Park, T., and Hubbard, E. E. 1997. Quality attributes and the selection of marketing channel by Georgia peach growers (Abstract). *American Journal of Agricultural Economics,* 79(5): 1738.

Hill, L. D. 1996. Quality in agricultural products, in *Quality of U.S. Agricultural Products,* L. D. Hill, ed., Task Force Report No. 126, Council for Agricultural Science and Technology, pp. 13–18.

Huang, Chung-Liang. 1979. Estimating U.S. demand for meat with a flexible functional form. *Southern Journal of Agricultural Economics,* 11(2): 17–20.

Jordan, J. L., Shewfelt, R. L., Prussia, S. E., and Hurst, W. C. 1985. Estimating implicit marginal prices of quality characteristics of tomatoes. *Southern Journal of Agricultural Economics,* 139–146, December.

Kriström, B. 1990. Valuing environmental benefits using the contingent valuation methods: An econometric analysis. *Umeå Economic Studies* No. 219, Univ. of Umeå, 170 p.

Lai, Y., Florkowski, W. J., Huang, C. L., Brückner, B., and Schonhof, I. 1997. Consumer Willingness to Pay for Improved Attributes of Fresh Vegetables: A Comparison Between Atlanta and Berlin. *Annual Conference of the Western Agricultural Economics Association*, Sparks, NV, USA, July 13–16.

Lancaster, K. 1966. A new approach to consumer theory. *Journal of Political Economy*, 74: 132–157.

McFadden, D. 1981. Econometric models of probabilistic choice, in *Structural Analysis of Discrete Data with Econometric Applications*, C. F. Manski and D. McFadden, eds., The MIT Press, Cambridge, MA, pp. 198–272.

Misra, S. K., Huang, C. L., and Ott, S. L. 1991. Consumer willingness to pay for pesticide-free fresh produce. *Western Journal of Agricultural Economics*, 16(2): 218–227.

Moteki, M., Muller, H. G. 1992. Comparative study of vegetable intake in Japan and the United Kingdom. *Journal of Consumer Studies and Home Economics*, 16: 317–329.

Nichols, J. P. 1996. Fruits and vegetables, in *Quality of U.S. Agricultural Products*, L. D. Hill, ed., Task Force Report No. 126, Council for Agricultural Science and Technology, pp. 139–177.

Park, T. and Florkowski, W. J. 1999. Demand and quality uncertainty in pecan purchasing decisions. *J. Agric. Applied Econ.*, 30(2): (forthcoming).

Rosen, S. 1974. Hedonic prices and implicit markets: Product differentiation in pure competition. *Journal of Political Economy*, 82: 34–35.

Shewfelt, R. L., Erickson, M. E., Hung, Y-C., and Malundo, T. M. M. 1997. Applying quality concepts in frozen food development. *Food Technol.*, 51(2): 56–59.

Theil, H. 1952. Qualities, prices, and budget inquiries. *Review of Economics and Statistics*, 19: 129–147.

Wahl, T. I., Shi, H., and Mittelhammer, R. C. 1995. A hedonic price analysis of quality characteristics of Japanese Wagyu beef. *Agribusiness*, 11: 35–44.

Waugh, F. V. 1928. Quality factors influencing vegetable prices. *Journal of Farm Economics*, 19: 185–196.

CHAPTER 13

Integrated Quality Management Applied to the Processed-Vegetables Industry

JACQUES VIAENE
XAVIER GELLYNCK
WIM VERBEKE

INTRODUCTION

A positive product perception by consumers is the base for increasing consumption, while a negative perception results in passive consumer behavior. The research question in this chapter centers on potential actions and communications to stimulate the consumption of processed vegetables. The research project is concerned with identifying the key factors that create the *competitive advantage* for processed vegetables and with formulating appropriate strategies to ensure that these advantages within the agri-food chain are fully exploited in the future.

The aim of the research is twofold: (1) to identify topics within Integrated Quality Management (IQM) in the processed-vegetables sector that correspond with consumer requirements, and (2) to develop communication about these topics within the IQM system to the consumer, focusing on competitive advantage.

This chapter emphasizes the state of the art related to quality management. Recent introductions of several relatively new concepts in the field of management will be presented, namely, Supply Chain Management, Efficient Consumer Response, Total Quality Management and Integrated Quality Management. The drivers of these changes in management will be described focusing on the creation of competitive advantage. A case study on quality management in the processed-vegetables industry will then be introduced. After discussing the research

246

methodology, the results of the project are presented. Finally, some conclusions and needs for future research are drawn.

STATE OF THE ART

To succeed in today's competitive agri-food marketplace, two options are available (Grunert, 1996) (Figure 13.1):

- organize production more efficiently
- meet consumer requirements

The organization of more efficient production relates not only to the individual company, but involves all links in the market chain from the supplier of raw materials to the farmer and to the final consumer. It concerns a strategy that focuses on cost saving. In European literature this concept is called Supply Chain Management (SCM). The SCM concept is of particular interest for logistic control (Evans et al., 1993; Handersson, 1995; Harland, 1995). Supply Chain Management is defined as a concern with the linkages in the chain, from primary producer to final consumer, with the incentive of reducing the transaction costs incurred (S-Bridge, 1996; Wilson, 1996). SCM seeks to break down barriers between each of the units so as to achieve higher levels of service and sub-

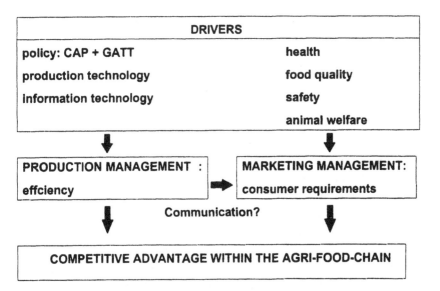

FIGURE 13.1 Competitive advantage and the agri-food chain.

stantial savings in costs. The concept of SCM is largely based on the Transaction Cost Economics (TCE) as a reaction to the neo-classical economic model, which suggests that all parties have the necessary information to be able to make rational choices within the exchange process (Loader, 1996). It is clear that this information is not (always) available in practice, a situation that creates transaction costs.

Economists largely ignored the original insights of Coase until the late 1960s and early 1970s, when the work of Williamson was published (Williamson, 1975, 1979; Williamson and Winter, 1993). Hereby, the cost-determining attributes of individual transactions are outlined as their frequency, the environmental uncertainty surrounding them and the specificity of the assets required to consummate them. However, there is still no clear definition of transaction costs (Loader, 1996). Major issues include:

- lack of definition of the concept of transaction costs (Williamson, 1979)
- the costs of running the economic system (Arrow, 1969)
- the resource inputs involved in transacting—defining, protecting, and enforcing the property rights to goods (North, 1989)
- resource losses due to lack of information (Dahlman, 1979)

Building relationships within the chain reduces uncertainty and transaction costs on the one hand and creates an access to economies of scale by bypassing traditional market arrangements on the other hand. In this way the cooperative behavior in relationships between channel members creates an opportunity for higher profits, attainable by the channel as a whole (Arndt, 1979).

In the U.S., Efficient Consumer Response (ECR) is the more common term for SCM. However, ECR goes further in that it combines both supply- and demand-side elements. ECR originated from discussions in the American food and beverage industry in 1992, which involved food producers, retail chains and industry organizations (FMI, 1993). The purpose of ECR is to increase the efficiency and effectiveness of the entire food chain by the integration of marketing and logistic decisions and optimal coordination between the different links throughout the chain. The ultimate goal is to maximize consumer satisfaction by a maximally performing chain (Corstjens and Corstjens, 1995; Buxbaum, 1995). It has been estimated that the application of ECR in the U.S. will lead to as much as $30 billion in total savings to the food industry (Van der Laan, 1994; Molpus, 1994).

ECR is often considered as a combination of demand-side management, also called category management, and supply-side management. The theory of ECR consists of four strategies regarding the marketing and logistics processes (Wierenga, 1997):

- category management, which refers to the processes that involve managing product categories as business units
- efficient replenishment, which aims at bringing the right product at the right time to the right place in the most efficient way
- efficient promotions, which relate to the efficiency of sales promotions to retailers and consumers and focus on avoiding excessive stocks and high performing production planning processes
- efficient product introductions, which aim at reducing the chance of failure through intensive collaboration between retailers and producers

Another concept focusing on the links throughout the chain is called Value-Adding Partnerships (VAP). The concept originated from contributions to the Functional School, which recognize that each enterprise has to perform specific functions within the whole of the processing and distributions processes between the primary producer and the final consumer (Alderson, 1957). Porter (1990) concretized these ideas within the concepts of the added value chain and the value system. A company's value chain consists of all activities such as production, marketing, delivery, service and supporting activities that contribute value to the buyer. Placed within an industry, this value chain is embedded in a larger stream of activities that together form the value system. It includes suppliers and the different stages in the distribution channel as well as the ultimate consumer.

Historically, companies at different stages of the system tended to behave in an autonomous way and demonstrated adversary, rather than cooperative behavior toward each other. However, it is increasingly apparent that the overall performance of the system can benefit from internal cooperation and partnerships. All participants benefit from the improved performance of the entire system.

Winning customers from competitors cannot only be realized by saving costs, but also by delivering greater value. A product offering good value creates consumer satisfaction. The complex idea of *quality* is closely related to satisfaction (Kotler and Armstrong, 1991). Quality has a direct impact on product performance and hence on customer satisfaction. Quality begins with customer needs and ends with customer satisfaction. While there are many definitions of quality, all share the common assumption that quality is determined by the customer (Cortada, 1993), and hence should be defined from a consumer-oriented perspective. Quality definitions of this kind include the following aspects:

- continuous improvement (Deming, 1986)
- fitness for use (Juran, 1989)
- conformance to requirements (Crosby, 1979)
- a product that is most economical, most useful and always satisfactory to the consumer (Ishikawa, 1985)

In recent years, many companies have adopted Total Quality Management (TQM) programs designed to constantly improve the quality of their products, services and marketing processes. TQM relates to all the processes in the organization that contribute directly or indirectly to delivering quality as defined by the consumer (Ross, 1993). The control component (quality assurance) has shifted from product inspection to process control. As a result, the processes in the various divisions or links of a company and the intermediate products and services are continuously improved in quality. It concerns the acceptance of TQM as a way of company life, by including all functions of the business and its integration into the product life cycle such as design, planning, control, production, distribution and field service.

TQM in each link and company of the market chain, and the coordination of these links, is called Integrated Quality Management (IQM). The concept of IQM goes beyond the individual company's quality management. It is referred to as "integrated" since the TQM concepts of individual companies in the chain are integrated in order to make them fit each other. The knowledge about and control of the production process and the coordination of all links is essential for a good quality and improved performance of the chain. It means that several of the above defined concepts are included in the concept of IQM, namely, TQM, SCM and ECR. The overall objective is to add value to the entire chain and to realize competitive advantages and a better performance of the entire chain.

DRIVERS OF COMPETITIVE ADVANTAGE

Trends or drivers create major economic opportunities for companies that demonstrate the ability to manage the system components to gain competitive advantage (Downey, 1996). As indicated in Figure 13.1, competitive advantage derives from the value a company offers to its buyers, on the one hand (marketing management), and from the costs incurred in delivering this value, on the other hand (production management).

The impact of *consumer values* and *preferences* is the single greatest driving force changing the structure of the vegetable chain from producer to consumer:

- An increasing health-conscious public has placed far greater importance on food quality, which makes health-related characteristics a strategic issue in building a competitive position in the market place.
- The increasing sensitivity of consumers to the safety and environmental aspects of agricultural products results in huge investments to build identity, brand recognition and public trust. Food companies attempt to build differential advantage upon the unique qualities of their products. However, the consumer is also used to an industry structure that boosts production efficiency and lowers food costs. Chains are challenged to design a framework that balances both consumer interests.

It is vital for the processing vegetable chain to transmit these consumer preferences throughout all stages of the vertical system. Therefore, it is necessary to gain insights into the way costs and benefits along the various stages of the chain are influenced and distributed. These insights may help to develop effective transfer pricing instruments. It is clear that in an economic environment prices are the efficacious incentives in affecting economic decisions to realize the transmission process (Viaene and Truyen, 1995).

Second, food and agricultural *policy* continue to play a major role in the development of the food and agribusiness chain. The reduction of agricultural support programs in the EU continues to place greater pressure for efficiency on the production sector. Moreover, the generally accepted scenario of EU enlargement and the coming World Trade Organization (WTO) negotiations are major turning points for the EU Common Agricultural Policy (CAP). This evolution will take place against a background of world markets becoming more volatile and competitive and will result in structural changes in food supply throughout the EU. The traditional family-oriented farms continue to be important, but the critical decisions about variety, price, production period and quality are part of a more sophisticated business approach. The policy aspects are also related to regulations with respect to product standards, environmental controls, pesticides and additives usage, which can result in competitive advantages for parties that are not affected or that have the know-how to deal effectively with the new situations.

Third, *technological* progress has created new opportunities in every part of the food chain. For example, advances in production technology allow farmers to increase both the efficiency and precision of the raw agricultural products supplied into the system. Biotechnology and genetic engineering are creating new products and new production processes in the food domain. In the nonfood domain, technologies are developed to use agricultural products as raw materials for industrial processes such as natural fibers and starch-based packaging materials.

Fourth, *information technology* creates new possibilities for management and control systems. The ability to measure more precisely and track product and processes more easily increases the obligation of every company to be responsible for its contribution to the final product.

The four groups of drivers result in more efficient production management on the one hand and a better integrated quality management on the other hand. The competitive position of the companies is improving by decreasing relative costs and increasing the perceived value for the consumer. The question remains how to communicate both advances to consumers. Corresponding with the IQM approach to vegetables, a communication strategy for consumers is developed.

RESEARCH METHODOLOGY: IQM FOR PROCESSED VEGETABLES

This research methodology is based on primary exploratory and conclusive research. Qualitative research by means of focus group discussions was established to gain preliminary insights into consumer attitude, perception and behavior toward vegetable consumption. Additionally, insights were gained about requirements of consumer information (chain perception) concerning the vegetable chain and potential topics for communication. Five focus group discussions with each six to eight respondents were conducted during March 1997. The participants were consumers of fresh and processed vegetables several times per month. Based on a topic list or interview guide, the moderator coordinated the discussion. For three to four hours the moderator applied the funnel approach and projective techniques to guide the discussion and probe the respondents to elicit insights. Next, based on the information gathered, hypotheses and key attention topics for quantitative conclusive research were drawn.

Second, quantitative primary data were gathered through a sample survey research. The research approach began with administering pretested formal questionnaires during personal in-home interviews led by trained field workers. The questionnaire comprised issues such as general consumer behavior and attitude toward vegetable consumption and consumers' chain perception of and interest in different processes within the processed-vegetable chain. The target population of the survey consisted of people living in Belgium, aged between 15 and 65 years, who were the main person responsible for purchasing vegetables within their household. The respondents were selected by means of nonprobability quota sampling. Quotas toward age and place of living were established. The overall sample size was set at $N = 500$ respondents, equally split up between Flanders (northern Belgium) and Wallonia (southern Belgium). This sample size satisfied the minimum sample size rules suggested by Sudman (1976). The fieldwork was conducted during October and November 1997. After coding and editing of the questionnaires, the collected data were analyzed by means of the statistical package SPSS.

RESULTS

The discussion of the results is presented in four major parts, combining in each part the results of the qualitative research (focus group discussions) and the results of the quantitative research (survey):

- consumer attitude toward vegetable consumption
- overview of consumer perception of fresh, canned, glass and frozen vegetables
- the ideal product identified from a consumer point of view
- IQM in the processed-vegetable chain and communication to consumers

Consumer Attitude Toward Vegetable Consumption

Based on the focus group sessions, it became clear that the consumer is confronted with a dilemma. Fresh vegetables are clearly considered as the ideal product but in some circumstances, such as unexpected situations (getting home late, visitors) and lack of time, preparation of fresh vegetables is impossible. Under these circumstances, the consumer looks for an alternative in frozen vegetables, canned vegetables or vegetables in glass. Focusing on vegetables in glass, three functions can be identified:

1. Ideal vegetable for fast and cold preparations such as celery and grated carrots
2. Useful in the case of unexpected situations such as coming home late and unexpected visitors
3. Alternative for fresh vegetables, of which the preparation is time-consuming, such as salsifies, asparagus, red cabbage, carrots and peas

An interesting topic resulting from the analysis of the purchase criteria for vegetables in glass concerns the fact that heavy users (several times per week) did not consider themselves as such. This attitude is related to a preference for the fresh vegetable and that housewives who only prepare and serve vegetables in glass fear being considered as less caring about their family and even as being lazy.

The attitude of the respondents toward vegetables in general was evaluated by means of a 3-point Likert (ordinal) scale that required the respondents to indicate a degree of agreement or disagreement with a series of statements derived from the qualitative research part. The profile analysis of the responses for the total sample is presented in Table 13.1.

Statistically significant differences in attitude between the four defined types of vegetable consumers were revealed by performing the Kruskal-Wallis nonmetric one-way analysis of variance. This test examines the difference in medians with the null hypothesis being that the medians of the four populations are equal. The significance of the computed chi-square statistic implies the rejection of the null hypothesis (Malhotra, 1996). The null hypothesis can be rejected for statement 1 ($p = 0.020$), 3 ($p = 0.029$), 4 ($p = 0.000$), 5 ($p = 0.010$), 6 ($p = 0.014$) and 9 ($p = 0.038$).

Consumers of fresh and frozen vegetables, when compared to consumers of vegetables in glass, agreed more with the statement that a meal is not complete without vegetables and with the statement that vegetables are the most important part of the daily meal. Consumers of fresh vegetables disagreed more than consumers of vegetables in glass that preparation is too time-consuming. The typical consumer of fresh vegetables also expressed both higher preference for vegetables produced within his/her country or region and a higher willingness to pay a premium price for environment-friendly produced vegetables. Consumers

Table 13.1. Profile Analysis for Evaluation of Attitude Toward Vegetables, % of Respondents (N = 500).

Statements	Agree	Neither Agree nor Disagree	Disagree	Total
1. A meal is not complete without vegetables.	85.4	5.0	9.6	100
2. Branded vegetables guarantee a better quality.	30.0	22.2	47.8	100
3. The preparation of vegetables is too time consuming.	27.0	11.8	61.2	100
4. Preferably, I buy vegetables from my own country or region.	44.4	21.0	34.6	100
5. Vegetables are the most important part of the daily dish.	59.8	18.8	21.4	100
6. I must see the vegetables to judge their quality.	86.0	7.8	6.2	100
7. Vegetables are more important in families with little children.	57.4	12.6	30.0	100
8. Preferably, I buy environment-friendly produced vegetables.	55.6	23.4	21.0	100
9. I am willing to pay a premium price for environment-friendly produced vegetables.	44.6	20.2	35.2	100

Source: primary research, survey.

of fresh vegetables also attached higher importance to seeing the vegetables as a guide to judging their quality.

PERCEPTION OF FRESH AND PROCESSED VEGETABLES

Within the focus group discussions, the perception of fresh, frozen, canned vegetables and vegetables in glass was determined by using a projective technique, namely, the photo sort. Hereby respondents are asked what type of consumer prefers fresh, frozen, canned vegetables and vegetables in glass. In this way, elements are collected enabling a description of the typical consumer of each category of products. In total, eight types of consumers were shown during the photo sort. The results of this exercise are shown in Figure 13.2.

The typical consumer of fresh vegetables can be described as a female who corresponds with the traditional mother figure and concentrates on healthy food, containing all minerals, vitamins and energy. Cocooning is a central lifestyle in the life of this person, who is positioned in the second quadrant of the figure.

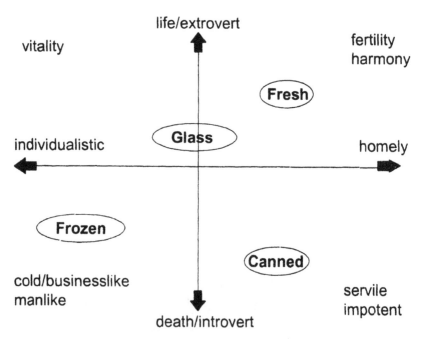

FIGURE 13.2 Perceptions of consumers of fresh, frozen, canned and glass-packaged vegetables.

The consumer of frozen vegetables is defined as someone who wants to serve fresh vegetables but looks for the ideal compromise, which is found in frozen vegetables. This type of consumer is described as modern, sophisticated and chic—a busy person, working outside the home, who considers cooking a waste of time. In the family of the consumer of frozen vegetables, the microwave oven takes a central place. In the figure, the typical consumer of frozen vegetables is situated in the third quadrant.

The typical consumer of canned vegetables is considered as old-fashioned, having no taste. This person is considered as introverted and servile. A typical characteristic of this person is a lack of spending much money on food. The typical consumer of canned vegetables is situated in the fourth quadrant of the figure.

The consumer of vegetables in glass occupies a dualistic position in the upper half of the figure. This person is perceived as active, working outside the home and responsible for a family. Normally, fresh vegetables are preferred, but the typical consumer of vegetables in glass does not like cooking and values personal time. Under these circumstances, vegetables in glass are an ideal alternative. Quality occupies a dominant position related to food choice. The fact that the core product can be seen through the glass helps to evaluate the quality. Another typical characteristic is related to the ideal that vegetables in glass give the possibility of meeting every wish in the family.

In quantitative research, the perception of different types of vegetables is assessed using a pick-any scaling technique. The pick-any scaling method differs from the more classical scaling methods such as the Likert scale or semantic differential in that respondents need less time and find the method easier, without any loss of reliability or validity (Van Kenhove, 1995). The respondents were asked to choose for each type of vegetable the most relevant attributes out of a list of attributes, resulting from the qualitative research. The selection of an attribute by the respondents means that the attribute is highly associated with the product discussed.

The attributes were split up in core product attributes, augmented product attributes, product benefits, situation factors and image components. The response profiles for each type of vegetables are presented in Table 13.2.

Fresh vegetables were most associated with the attributes healthy, tasty and natural. The perceived benefit of fresh vegetables is that they provide people with the necessary vitamins and minerals. Fresh vegeta-

Table 13.2. *Perception of Vegetables on Five Attribute Lists, % of Respondents (N = 500).*

	Fresh	Frozen	Canned	Glass
Core Product				
Has a good taste	29.1	11.1	6.5	7.2
Contains a lot of minerals and vitamins	20.0	23.3	5.9	10.9
Is cheap	3.4	6.4	29.7	13.3
Is healthy	35.1	6.4	2.6	4.0
Has a good quality	11.4	12.7	8.9	11.9
Is easy to prepare	1.0	40.0	46.5	52.7
Augmented Product				
Natural	73.2	24.2	6.4	9.3
Traditional	5.0	21.8	13.0	20.5
Organic	12.8	4.8	3.0	3.6
Industrial	1.0	32.1	50.9	37.0
Forced	1.0	9.5	15.0	10.7
Old-fashioned	1.5	2.6	9.8	9.9
Environment-friendly	5.8	5.0	1.8	9.1
Product Benefit				
Provides necessary minerals and vitamins	68.4	21.9	6.7	7.4
Presents well	6.6	14.3	6.7	22.7
A taste for everybody	4.2	21.3	35.2	17.7
Quality can be judged by seeing	10.4	5.2	1.8	23.3

Table 13.2. (continued)

	Fresh	Frozen	Canned	Glass
Is good for the environment	6.0	4.2	0.6	4.0
Provides variety	4.4	33.0	49.1	24.7
Situation Factor				
For fast preparation	3.0	39.2	47.4	45.0
For cold dishes	9.8	0	6.6	16.1
For dishes prepared for the whole family	58.0	8.8	5.6	4.6
For summer time	13.2	1.0	1.8	3.4
For storage purposes	4.8	28.7	24.9	16.3
For winter time	1.0	16.9	8.6	9.0
For dishes prepared for myself	10.2	5.2	5.0	5.6
Image Component				
For the real family mother	36.8	9.7	5.0	5.5
For active people	13.6	32.2	25.4	25.9
For connoisseurs	24.0	4.4	3.0	1.8
For cheerful people	4.4	3.4	2.4	3.2
For modern people	3.6	27.2	22.3	23.4
For old-fashioned people	1.8	2.2	8.7	10.1
For people with little children	12.8	4.6	3.6	3.6
For single people	3.0	16.3	29.6	26.5

Source: primary research, survey.

bles are served in meals prepared for the whole family and most appreciated by the real family mother and connoisseurs. No statistically significant differences were found in the perception of fresh vegetables between the four groups of vegetable consumers.

Of all processed vegetables, frozen vegetables had the best image in terms of vitamin and mineral content. This attribute was significantly more mentioned by typical consumers of frozen vegetables. Frozen vegetables were perceived as easy and fast to prepare, providing variety and ideal for active and modern people. An association with the attribute "industrial" was mentioned. Canned vegetables were perceived as cheap and industrial. Other associations included convenience, speed and variety. Canned vegetables had the image of being the preferred vegetable of single people. Glass vegetables had a less industrial and cheap connotation but a more traditional image than canned vegetables. Apart from offering variety, important product benefits included a good presentation and that quality of the product can be judged visually. A similar response pattern for glass and canned was found for situation and image components.

These quantitative results about perception and image of vegetables fully confirm the findings of the qualitative research. That is, frozen vegetables are clearly the preferred alternative for fresh, which can be explained by greater familiarization of consumers with the freezing preservation technique rather than with the vegetable sterilization technique, which was commonly applied in households some decades ago.

The Ideal Product

A definition of the ideal product is determined on the basis of asking the consumers to describe the ideal production process from the beginning till the final prepared product as it is presented on a plate. During the focus group discussions, respondents were asked to think about every step, each element in the production process and to describe the ideal picture. Qualitative research revealed that consumers attached specific attention to the soil, the seed, the growth process, harvesting practices, vegetable processing and preparation. Using a pick-any scale enabled a quantitative assessment of the consumer concerns related to each of these six steps in the production process. For each step in the process, relevant attributes were selected based on the qualitative research. These attributes were labeled as "must" and "should not" be present, which, respectively, means that the ideal production process "must" take care of specific practices and "avoid" others. The attributes were presented to the respondents, who were asked to indicate the most relevant attributes for their imagined ideal vegetable production process.

For each step in the production chain, an importance weight coefficient was calculated. An initial response to the results was that the differences in importance attached to chain processes were rather small. All steps were given similar importance by the consumers. Nevertheless, it was perceived that greatest importance was associated with those processes with which the consumer had the greatest familiarity through their own experiences in vegetable production: tillage or soil cultivation, harvesting and vegetable preparation.

With respect to the soil, consumers stressed that the soil must be pure and carefully cultivated. Treatments with pesticides and, to a lesser extent, chemical fertilizers were not accepted for the ideal vegetable production process. Soil purity was statistically more likely to be mentioned by typical consumers of glass and canned vegetables than by others. The seed for the ideal vegetable production must be of premium quality and should not be treated or coated with pesticides. In addition, about a quarter of the respondents indicated that the seed should not be genetically engineered. The growth process must be under permanent control. Again reservations about the use of pesticides were the major concern, together with the rejection of any kind of sprays on the vegetables. The harvesting process must be conducted at the right moment, especially not too late. Care should also be given to avoid damaging the vegetables during the harvest.

During processing, special attention must be paid to careful washing of the vegetables and to strictly minimize the storage period between harvesting and processing. Consumers of canned and glass vegetables attached significantly more importance to "not supplying vitamins" during the processing of the vegetables. During preparation, loss of taste, vitamins and minerals should be avoided. This topic was significantly more stressed by frozen and fresh-vegetable consumers. Finally, all family members must appreciate the ideal vegetable as it is served.

The identified topics from a consumer viewpoint for the ideal vegetable production chain reveal both key factors for successful improvement of the processes in the processed-vegetable chain and relevant topics for communication with consumers around chains. It is obvious that some topics related to the production process, such as manual labor or avoiding the use of machinery cannot be realized in practice. A great majority of the indicated issues for ideal vegetable production were, however, perfectly feasible in practice.

Concerns about the use the pesticides confirm the findings of previous research by Dittus and Hillers (1993, 1996). Their consumer research about pesticide use in vegetable production led to the conclusions that concerns dealt with residue effects on both personal health and the environment. They also revealed that concerns about pesticide residues on vegetables might confound appreciation of the nutritional merit of these foods. In our survey, concerns about pesticide were emphasized, which indicates that these practices should be approached with caution. Any technological improvement within the vegetable chain should be communicated to the consumer.

Several topics related to the production process of vegetables in glass, such as sowing seeds by children and manual removal of weeds, are not practical. However, these suggestions provide important topics for communication with consumers.

IQM and Communication

A communication plan typically includes the following elements (Kotler & Armstrong, 1991):

- identification of the target audience and quantitative objectives in terms of market share
- formulation of the message (content, structure, format)
- choice of the media (promotion above-below the line)
- definition of the source, namely, who is sending the message
- collection of feedback

Once the elements of the communication plan are determined, the developed communication concept must be tested (e.g., through indepth interviews). Through this test it is verified whether or not the message is understood and if the medium and source are accepted, liked and trustworthy.

Related to the stipulated objective, two approaches of IQM and communication can be advanced. It is important to mention that the discussion focuses on the way the consumer perceives IQM and more specifically the way vegetables are processed. For companies willing to defend their status as market leader or premium brand, the whole concept of IQM focuses on providing the "the ideal product" as a kind of guarantee concept. These elements offer a tremendous opportunity for chains that manage first to respond to these consumer preoccupations and second to use them as an effective communication tool. Alternatively, focusing on a specific market segment creates the opportunity to increase market share. It is possible to identify differences between groups of consumers (typical consumers of fresh, frozen, canned and vegetables in glass) in terms of perception of the ideal product. By focusing on these differences, a chain must be able to develop a communication strategy aimed at gaining market share by winning specific consumers at the expense of another type of processed vegetables.

In developing communication about IQM for processed vegetables, the following topics should be kept in mind:

- Consumers have a poor knowledge about the processing technique related to vegetables in glass, more specifically the sterilization process. It relates to the fact that consumers are much more familiar with freezing rather than with sterilization.
- By communicating with consumers about vegetables, nonscientific language should be used. It relates to the emotional product approach and the fact that a scientific language could be experienced as "chemical."
- Within the communication process, children and the farmer should retain a central place as symbols of purity, harmony with nature and future.

CONCLUSIONS AND RESEARCH AGENDA

During the last 10 to 15 years, new management techniques such as supply chain management, efficient consumer response, value-added partnerships, Total Quality Management and Integrated Quality Management have been introduced in the agri-food business. These techniques focus either on production management, on marketing management or both. Changes in the working conditions of the agri-food business, such as the desire of consumers to know the origins and production processes of the products they buy, the need of retailers to increase the efficiency of the channel, and the need of the primary producers to be assured of a destination for their products, create tremendous opportunities for chains on their way to develop competitive advantages.

The evaluation of consumers' needs and interests to know the production process and their ideal image of vegetables reveals several opportunities. The vegetable consumer is confronted with a dilemma. The consumer approaches vegetable consumption emotionally. Fresh vegetables are perceived on the one hand as the best product in terms of health, quality, nutrition and naturalness. On the other hand, the preparation of fresh vegetables is increasingly considered as too time-consuming, especially by working people. Under these circumstances, the consumer looks for an alternative under the form of processed vegetables: frozen, canned or glass. To justify this choice to other people such as family members, the consumer looks for rational support. Much of this rational support is identified by the elements provided during the description of the ideal product in terms of the ideal vegetable production process. Consumers' search for rational support offers great opportunities for chains that manage, first, to guarantee integrated quality and chain management, and second, to work out the realized consumer driven chain improvements as an effective communication tool. It is up to the processed-vegetables chain to translate these opportunities into adapted processes and institutional adjustments in order to realize and implement the consumers' ideal chain perception.

This chapter focused on Integrated Quality Management in agri-food chains, with application to the processed-vegetable chain. The research illustrates how theoretical concepts can be applied in practice through consumer research. The basic idea is to consider consumer satisfaction as the ultimate goal of IQM. Hence, filling in IQM in practice is impossible without first questioning the actors at the final and ultimate level: consumers. Further integration of consumer research in chain and quality management deserves attention in the future. Such attention should focus both on methodological research and analysis aspects in empirical work as well as on the translation of findings to become useful in practice from the marketing and management perspective of agrifood chains.

REFERENCES

Alderson, W. 1957. *Marketing Behaviour and Executive Action,* Richard D. Irwin, Homewood, IL.

Arndt, J. 1979. The domestication of markets: From competitive markets to administered inter-organizational marketing systems, in *Contemporary Issues in Marketing Channels,* Lusch, R. F. and Zinszar, P. H., Eds. University of Oklahoma, Norman, pp. 55–61.

Arrow, K. J. 1969. The Organization of Economic Activity: Issues pertinent to the choice of market versus non-market allocation, in Joint Economic Committee,

The Analysis and Evaluation of Public Expenditure: The PPB System, Vol 1, US Washington, DC: Government Printing Office, pp. 59–73.

Buxbaum, P. A. 1995. Some food for thought. *Distribution*, 94(1), 12.

Corstjens, J. and Corstjens, M. 1995. *Store Wars: The Battle for Mindspace and Shelfspace*, Wiley, Chichester.

Cortada, J. W. 1993. *TQM for Sales and Marketing Management*, McGraw-Hill, Inc., New York.

Crosby, P. B. 1979. *Quality Is Free: The Art of Making Quality Certain*, McGraw-Hill, Inc., New York.

Dahlman, C. J. 1979. The problem of externality. *J. Law and Economics*, 22 (1): 141–162.

Dittus, K. and Hillers, V. 1993. Consumer trust and behavior related to pesticides. *Food Technology*, July, 87–89.

Dittus, K. M. and Hillers, V. N. 1996. Attitudes about the nutritional benefits and pesticide-exposure risks from fruit and vegetable consumption. *Family and Consumer Sciences Research Journal*, 4(2), June, 406–421.

Deming, W. E. 1986. *Out of the Crisis*, 2nd Ed., MIT Center for Advanced Engineering Study, Cambridge, MA.

Downey, W. D. 1996. The challenge of food and agri products supply chains, in *Proceedings of the 2nd International Conference on Chain Management in Agri- and Food Business*, Trienekens, J. H. and Zuurbier, P. J. P., eds., Department of Management Studies, Wageningen Agricultural University, May, pp. 3–13.

Evans, G. N., Naim, M. M., and Towill, D. R. 1993. Dynamic supply chain performance: assessing the impact of information systems. *J. Logistics Information Management*, 6(4): 15–25.

FMI, Food Marketing Institute. 1993. *Efficient Consumer Response*. Washington, DC.

Grunert, K. G. 1996. Research on Agri-Chain Competence and Consumer Behaviour. Presented at EU-Workshop on Agri-Chain Competence: Learning from Other Chains. 's-Hertogenbosch, The Netherlands, March, 23 p.

Handersson, H. 1995. *Developing Relationships in Business Networks*, Routledge, London.

Harland, C. 1995. The dynamics of customer dissatisfaction in supply chains. *Int. J. Production Planning Control*, 6(3): 209–217.

Ishikawa, K. 1985. *What Is Total Quality Control? The Japanese Way*. Translated by Lu, D. J., Englewood Cliffs, Prentice Hall, NJ.

Juran, J. M. 1989. *Juran on Leadership for Quality: An Executive Handbook*, Free Press, New York.

Kotler, P. and Armstrong, G. 1991. *Principles of Marketing*, 5th ed., Prentice-Hall International Editions, Englewood-Cliffs, NJ.

Loader, R. 1996. Transaction costs and relationships in agri-food systems, in *Proceedings of the 2nd International Conference on Chain Management in Agri-*

and Food Business, Trienekens, J. H. and Zuurbier, P. J. P., eds., Department of Management Studies, Wageningen Agricultural University, May, pp. 417–429.

Malhotra, N. 1996. *Marketing Research: An Applied Orientation,* Prentice Hall, Upper Saddle River, NJ.

Molpus, C. M. 1994. Variety, not duplication. *J. Progressive Grocer,* (Jan), 31–32.

North, D. C. 1989. Institutions and economic growth: A historical introduction. *J. World Development,* 17(9): 1319–1332.

Porter, M. 1990. *The Competitive Advantage of Nations,* McMillan Press, London.

Ross, J. E. 1993. *Total Quality Management: text, cases and readings,* St-Lucie Press, Delray Beach, FL.

S-Bridge, D. 1996. Supply chain management for fresh vegetables: The key of success factors from the producer's point of view. *J. Farm Management,* 9(7), Autumn, 357–365.

Sudman, S. 1976. *Applied Sampling.* Academic Press, New York.

Van der Laan, A. 1994. Categorie management en ECR leveren 1.7 miljard winst op. *J. Food Personality,* (Sept): 8–9.

Van Kenhove, P. 1995. A comparison between the "pick any" method of scaling and the semantic differential. Working Paper 95/10, Department of Marketing, University of Ghent, 14 p.

Viaene, J. and Truyen, A. 1995. Naar kwaliteitszorg in de Belgische braadkippensector. Working Paper, Division Agro-Marketing, University of Ghent, March, 15 p.

Wierenga, B. 1997. Competing for the future in the agricultural and food channel, in *Agricultural Marketing and Consumer Behavior in a Changing World,* Wierenga, B. et al., eds., Kluwer Academic Publishers, Norwell, MA, pp. 31–55.

Williamson, O. E. 1975. *Markets and Hierarchies: Analysis and Antitrust Implications. A Study in the Economics of Internal Organization,* The Free Press, New York.

Williamson, O. E. 1979. Transaction-cost economics: The governance of contractual relations. *J. Law and Economics,* 22: 233–262.

Williamson, O. E. and Winter, S. 1993. *The Nature of the Firm Origins, Evolution and Development,* Oxford University Press, New York.

Wilson, N. 1996. The supply chains of perishable products in northern Europe. *British Food Journal,* MCB University Press, 98/6: pp. 9–15.

CHAPTER 14

Methods and Examples of Integration

STANLEY E. PRUSSIA

PRINCIPLES

AN integrated view of fruit and vegetable quality requires a systems approach for reducing losses and improving consumer satisfaction. In general, systems are formed by integrating components into a functioning whole. The components of a system cannot be understood without considering the system they help to form. Research efforts to improve a system have the purpose of changing existing components or designing new ones. Such changes require an understanding of the new interactions that will result among components of the changed system and interactions with other systems.

This chapter uses visual models of systems to clarify the complex interactions of components and systems necessary for providing consumers with affordable food that has desirable attributes. The process of making diagrams draws attention to the flows of product, money, and information from one system to others. Diagrams also help to visualize the integration of businesses producing and handling products. Established principles are used to determine what properties are necessary for a group of components to be viewed as a system.

Applications of the Soft Systems Methodology (SSM) are presented to show ways to include the human component in the systems that are necessary for providing fruits and vegetables to the final consumers. Systems with humans are considered soft because each person can have a

267

different viewpoint of a situation. SSM provides the systems thinking necessary to help those involved define existing situations, identify the changes desired, and take actions to improve.

LITERATURE

Models help communicate concepts, understand interrelations, and develop new insights (Wilson and Morren, 1990, p. 75). Physical models of an object or system are typically made to change the scale, represent conditions at a point of time, or simulate output for varying conditions. Models of airplanes, for example, are used in wind tunnels to study aerodynamics. Photographs represent a person's appearance at the moment the shutter is snapped. Designers often use drawings to stimulate changes while developing innovative solutions.

Symbolic models make it possible to communicate ideas and concepts about how things function. Simple x-y plots help to visualize trends, data variability, and other statistical information. Mathematical equations are symbolic models ranging from simple linear expressions to complex representations of nonlinear, time-varying, stochastic systems. Schematic diagrams are also symbolic models commonly used for electrical circuits, distribution of materials, organization responsibilities, and other visual displays. A key advantage of diagrams is the ability of users to process information in parallel rather than in a series as when reading words.

Visual models for showing systems are based on drawing a boundary to show those components inside the boundary that constitute the system and those outside the boundary (Wilson, 1990). The following is a summary of the properties listed by Wilson and Morren (1990, p. 190) as required for a system to qualify as a system:

1. Has an ongoing purpose.
2. Has a measure of performance.
3. Has a process of decision making and resource allocation.
4. Has components (subsystems) that exhibit all the properties of a system.
5. Has components that interact.
6. Exists within a wider environment with which it interacts.
7. Can be distinguished from the environment in which it exists by a boundary that represents the interface between the system and its environment.

8. Has both physical and abstract resources at its disposal.
9. Has some guarantee of continuity or stability.

The importance of taking a systems approach to improving fruit and vegetable quality was emphasized by Shewfelt and Prussia (1993), who highlighted the need to consider preharvest factors, include management functions, and emphasize latent damage (damage caused at one point but not detected until later). While preparing their book they realized that postharvest systems did not exist. Rather, a more realistic model was to view postharvest businesses as links in a chain.

The Soft Systems Methodology (SSM) was developed recently at the University of Lancaster, England (Checkland, 1993; Checkland and Scholes, 1990), for understanding and improving the ill-structured, messy problem situations found in the real world. A strength of SSM is its ability to combine technical issues with social and managerial issues found in what are called human activity systems. As mentioned, the term "soft" is used because each person viewing a real system may have a different understanding of its purpose, goal, and related problems. SSM is commonly described as a seven-stage approach to working on undefined problem situations to bring about improvements. The stages are as follows:

1. Identifies a situation considered problematic.
2. Expresses the elements of structure and process and the climate surrounding them.
3. In this stage the effort moves from the real world to systems thinking about the real world by writing "root definitions" (RD) that describe several viewpoints of the problem situation.
4. Develops a conceptual model (CM) for each RD in the form of a diagram showing activities and interactions among them.
5. Compares the CM diagrams from stage 4 with the real world situation described in stage 2.
6. The differences found in stage 5 stimulate debate, which leads to the definition of feasible and desirable changes to the CM in stage 4.
7. Actions are identified and taken to improve the problem situation presented in stage 1.

The cycle can be repeated after the situation originally described in stage 1 has changed due to the process. The root definitions are evaluated for completeness by using the mnemonic, CATWOE, where C is Customer, A is Actors, T is Transformation, W is Weltanschauung or Worldview, O is Owner, and E is Environment.

Conceptual models typically contain five to nine verbs, which represent what activities are the minimum necessary for them to function as the system described in the root definition. Normally, one or more of the activities is expanded into a hierarchy of systems by asking how the activity is accomplished. The resulting subsystem consists of verbs describing how one of the activities in the system is carried out. Thus, the "hows" for an activity in the system become the "whats" of the subsystem. Careful attention to a hierarchy helps to keep activities at similar levels.

STATE OF THE ART

Most verbal descriptions and diagrams for fruits and vegetables are based on a general grouping of all the related businesses, agencies, and other infrastructures into what is called the industry, the food sector, the agricultural marketing system, or the postharvest system. Very few efforts have been made to carefully describe the systems involved in terms of the properties needed for components to be a system. Visual models provide the opportunity to visualize systems in a hierarchy and to communicate other system ideas.

Most studies using SSM have been conducted to improve the organization that requested the study. Far less experience is available in applying SSM to situations where there are multiple businesses. Early efforts (Prussia and Shewfelt, 1993) have shown the value of using SSM for integrating viewpoints on systems for fruit and vegetable production, handling, marketing, and consumption. Additional examples are provided in the following section.

EXAMPLES OF VISUAL MODELS

Most fruits and vegetables reach a final consumer after passing through several businesses that can be viewed as a chain. For example, consider a peach eaten by an office worker in New York City after it was selected from a bulk display and purchased at a local supermarket. The peach could have been held in a warehouse several days after a two-day journey by truck from a packinghouse in Georgia where it was grown on a family farm.

Links in a Chain

The chain in Figure 14.1 represents the businesses needed to deliver the peach from the tree in Georgia to the office worker in New York

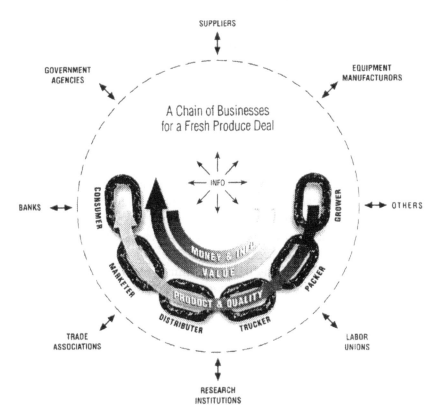

FIGURE 14.1 Businesses are represented as links in a chain for a fresh produce deal. Intensity changes along an arrow indicate increasing value or decreasing magnitudes of money, product, or quality. Information flows from consumer to grower, among links, and between links and supporting organizations. The circle is dotted to emphasize that the chain is not a system with a defined boundary.

City. Each link of the chain in this example is shown as a separate business.

Chains

Other chains could have supplied our office worker in New York City with the same peach picked the same day at the same farm in Georgia. The peach could have been purchased from a street vendor who contracts for fruit from a wholesale dealer operating a warehouse at the Hunt's Point Terminal Market. The truck could have had air suspension rather than metal springs. The container could have included packing trays to form layers of fruit rather than being packed in bulk.

When we expand the chain analogy we quickly realize that a multitude of chains are needed to supply other fresh fruits and vegetables to consumers in other locations. Frequently, a portion of the chain shown in Figure 1 is called a "deal," which exists for only a few weeks between a buyer and a packer. Even during a deal there are different chains because different transportation companies could have been used. Other chains would be required to provide peaches grown in California or in Chile when peaches are not available from North America.

In reality, each fruit or vegetable that is consumed has been subjected to a unique combination of inputs that affects each quality attribute of that item. Using peaches again for an example, each fruit on the same tree receives various amounts of sunlight and nutrient inputs and is picked at different maturities. After harvest each peach receives different physical and environmental stresses. The fruit that hits the bottom of a pallet bin is more likely to have a bruise than later arrivals. Some fruit has longer delays before cooling than others.

There is a wide range of impact levels on fruit passing through the packing line. Likewise, peaches undergo a range of temperatures, impacts, relative humidities, gas environments, and other handling conditions during transport, distribution, marketing, and in the consumer's home. Consequently, each fruit reaching a consumer has a unique history that determines its quality attributes when eaten.

Product Quantity

Each business link in the food chain shown in Figure 1 has control over the conditions of handling only while the products are in their possession. The links are held together by the flow of product shown by the arrow extending from the grower to the consumer. The handling conditions at each link determine the amount of product that reaches the consumer. The decreasing intensity along the arrow indicates a decrease in both quantity and quality as product moves to the consumer.

Annual food losses in the U.S. are estimated at 27% (96 billion pounds of edible food) for only the retail, food service, and consumer links (Kantor et al., 1997). Additional losses occur at the distributer, trucker, packer, and grower links. Losses are especially critical for developing countries (Okezie, 1998). Fruits and vegetables have higher rates of loss than foods in general. All four shipments in a replicated study (Campbell, 1985) by the University of Georgia had measured losses in excess of 50% for tomatoes shipped from growers to retailers. Retail sales of fruits and vegetables in the U.S. in 1997 were $64.75 billion. A 30% loss represents $19.4 billion per year. A reduction in loss from 30 to 20% would save over six billion dollars annually.

Product Quality

Quality is also lost as produce moves from farm to consumer. Fruit and vegetables available for consumption often suffer from bruises, loss of color, wilting, decay, mold, and other defects resulting from preharvest and handling conditions. Declines in nutritional value are documented (Watada, 1987). However, there are two exceptions to the common statement that postharvest handling can only reduce the rate of quality loss, not improve the quality of produce. First, pears, bananas, avocados, and other climacteric fruit undergo physiological changes after harvest that increase consumer acceptability.

Secondly, the quality of produce leaving a packinghouse should be better than what entered due to culling of unacceptable items and sorting of acceptable items into common sizes, maturities, or other desirable characteristics. The issue is whether reference is made to the individual item to or the quality of the product in the box or shipment. (Sorting and separating by humans on a flow of product is a production process for improving quality, not an inspection operation for quality control.)

Value

As depicted in Figure 14.1, value increases along the chain from the grower to the consumer. Thus, product is worth more when it is at the point of geographical need. Value is also increased when the product is available at the time needed and in the quantities desired. The goal of businesses at each link is to add value to the product at a rate higher than the rate of decline in quality.

Money

Money from millions of consumers funnels to the supermarket cash register (till). The arrow for money in Figure 1 shows this flow from the consumer to the grower. The decreasing intensity indicates that the amount of money decreases as it flows along the links. Approximately 30% of the original amount reaches the grower for equipment, supplies, labor, taxes, and so on.

Information

Ideally, information flows along with the money from the consumer to the grower as indicated by the arrow. Unfortunately, desired specifications for firmness, color, size, shape, blemishes, and other quality attributes often do not filter through the businesses to the packer and grower. One reason is that different businesses place different weights or values on each quality attribute. For the peach example, the packer requires the grower to pick the fruit when it is less than fully ripe so that it will not be too soft when it arrives in New York City. Thus, the consumer receives a fruit that has less flavor than it would have if it remained on the tree a few extra days.

Figure 14.1 also shows information as multidirectional to indicate diverse flows. Some information flows along the chain from the grower/packer to the consumer such as brand labels. Other sources of information for the businesses in the chain include the organizations shown outside the dotted circle (trade associations, government agencies, research institutions, equipment manufacturers, suppliers, banks, labor unions, and many other groups and organizations).

System Boundaries

The chain of businesses shown in Figure 14.1 is commonly referred to as a postharvest system. Similar terms are used to describe banking systems, educational systems, transportation systems, and other national or international infrastructures. However, recent experience has demonstrated the value of visualizing the businesses shown inside the dotted circle of Figure 14.1 as links in a chain rather than as a system. The circle around the chain is dotted to indicate it is not a system boundary.

The chain in Figure 14.1 fails to satisfy properties 3, 7, and 9 of systems, as listed previously in the Literature section of this chapter. Returning to our earlier example for peaches, it is clear that a mechanism does not exist that is valid for regulating all the links (item 3). There is no decision-making function outside the chain that has the authority to allocate resources to all the links. Even the most vertically integrated supply chains (such as for bananas) do not have control at the retail store or have control over the consumer. No one person or entity owns all the links in a chain.

A solid circle representing a boundary (item 7) cannot be drawn to replace the dotted circle because there is no reason to exclude any of the organizations shown outside the dotted circle. If any number of components other than the links of the chain could be inside the boundary, usually called the postharvest system, then it would not be possible to distinguish between components inside or outside the arbitrary boundary. It will be shown that each link meets the requirements for being a system, including the ability to draw a boundary separating the component parts in the link from other links in the chain and from the organizations in the environment (suppliers, banks, etc.).

The wide range of possible chains presented in the peach example demonstrate that food chains frequently form and disperse. Thus, the requirement that a system have continuity (item 9) is not satisfied.

Recognizing that food chains are not systems helps to understand why it is difficult to make changes that promise to reduce losses or improve the quality of products available to consumers. A change typically cost money and there is no authority outside the chain that can require that higher profits for one link are transferred to another link that made the investment. Similarly, it is difficult to obtain research funding to improve the chain because there is not a funding source (other than some government agencies) that takes responsibility for all the links or more general priorities. Thus, research tends to focus on improving only one business link.

Links Are Systems

Each link shown in Figure 14.1 meets all nine of the requirements to be viewed as a system. Consider the consumer system (link).

1. The consumer has the ongoing purpose of selecting, purchasing, and consuming fresh fruit.
2. Performance of the system is measured by how satisfied the consumer (system owner) is with the availability, price, and quality of the peach.
3. Decisions are made on where to purchase, how much to pay, and level of satisfaction.
4. Subsystems for the consumer system include activities to obtain, store, prepare, and eat fruit along with the need to manage the system.
5. The subsystems interact.
6. The consumer interacts with a wider environment.

7. There is a clear boundary between what the consumer has control over and the other systems outside the consumer's boundary that can only be influenced by the consumer (retailer, research institutes, etc.).
8. The consumer has financial and other resources needed to purchase fruit.
9. The consumer expects to continue buying and eating fruit for many years.

The purpose of presenting the model in Figure 14.1 is to help visualize the interactions of existing systems for making one fresh produce item available to a final consumer. Multitudes of similar chains are required for all products, across wide geographical distances, and for diverse distribution channels.

The huge variety of fresh products available 365 days a year at the produce department of supermarkets demonstrates the tremendous success of previous efforts. However, there is considerable discontent by consumers with the quality of fresh produce. Flavor is often lacking, bruising can be excessive, and texture can be unacceptable. At the same time the losses mentioned earlier increase the costs of products to the final consumer.

Expanding a Link

The previous section presented a method for visually integrating the systems and flows necessary to provide an individual produce item to a final consumer. The method for visualizing the chain is now expanded by showing the subsystems for two of the links in the chain.

Two roundtable workshops were recently convened by the Postharvest Active Learning Laboratory (PAL Lab) at the University of Georgia. The 20 participants included managers from the links shown in Figure 14.1. Some participants stated that it was the first time they had met with representatives from each business to discuss how to reduce losses and to improve the quality of fresh produce. At the second roundtable we developed diagrams to show the activities required by each business link. The consumer link was done as a group before small teams worked on the other links.

The model for a *consumer* in Figure 14.2 is based on inputs from the roundtable participants. Five subsystems were identified as the primary activities of a consumer of fresh produce (manage the system, obtain fresh produce, store products obtained, prepare as necessary, and to eat

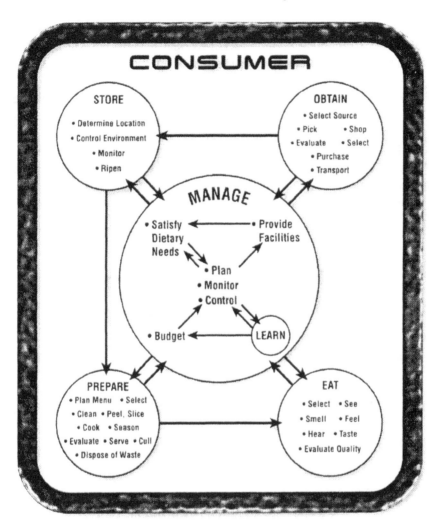

FIGURE 14.2 Expansion of the *consumer* link in the chain of businesses shown in Figure 14.1. The verbs are tasks that are performed, except for the *learn* activity, which is a sub-subsystem expanded to the task level in Figure 14.3.

the item). Each subsystem contains the tasks necessary for the subsystem to function. For example, the subsystem for *obtain* lists the tasks: select source, shop, evaluate, pick or select, purchase, and transport. The *manage* subsystem in Figure 14.2 shows *learn* as a sub-subsystems rather than a task. The activities and tasks are deliberately identified as verbs to indicate they are actually performed.

The *learn* sub-subsystem is shown in Figure 14.3 to demonstrate the task level. Combining Figures 14.1, 14.2, and 14.3 we see a progression from the complex interactions of national or international businesses and organizations to a list of tasks that are performed by an individual.

Small teams of roundtable participants then completed models for the other links in the chain. Figure 14.4 is based on their input for the *trucker* link. The team identified five activities needed for a business transporting fresh produce from the packer to the distributer (manage business, manage employees, provide services, supply equipment, and transfer product). Each subsystem includes activities that are tasks to be completed or a sub-subsystem that is modeled separately. The *transfer prod-*

FIGURE 14.3 The LEARN sub-subsystem from the *consumer* system shown in Figure 14.2 is expanded to show tasks that are performed.

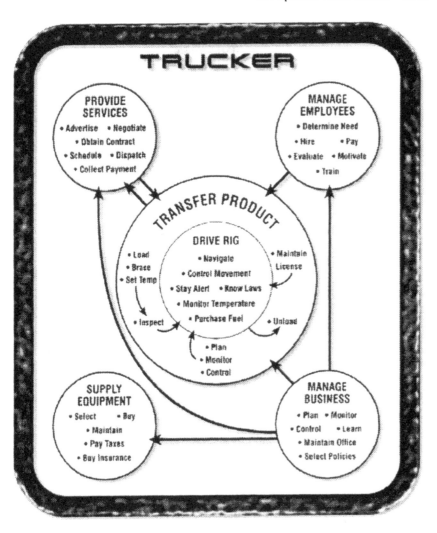

FIGURE 14.4 The *trucker* system from Figure 14.1 expanded to show tasks and the sub-subsystem *drive rig* with its own tasks.

uct subsystem includes the sub-subsystem *drive rig* in addition to the tasks shown.

The purpose of future roundtable workshops will be to evaluate the impact of changes at one link on subsystems in other links. For example, the retailer could pay more for shipments that had a higher percentage of inedible items removed by the packer because the value of product discarded at retail is about four times what it is at the packer.

The costs for discarded items at the packer link is about one fourth the value of the same item if discarded by the retailer or consumer. All businesses in the chain could benefit from not paying to cool, pack, transport, distribute, and retail items that are discarded by the *marketer* or *consumer*. The visual models will thus provide a tool for identifying realistic changes for improving the entire chain.

EXAMPLES OF SOFT SYSTEMS METHODOLOGY

Applications of a Soft Systems Methodology (SSM) by the author to fresh fruit and vegetable systems are described in the following sections.

England

Two British companies that specialize in fresh produce importing agreed to participate in a study that included using SSM for describing their operations. The objectives of the study by Prussia and Hubbert (1992) were to: (1) learn the operations of fresh produce importing companies and (2) evaluate the applicability of SSM for modeling multibusiness postharvest systems.

Both companies were major suppliers to British supermarket chain stores. The corporate directors and other top-level managers in both companies were interviewed. They identified five to seven of the main activities they controlled by giving verbs representing each activity. Root Definitions (RD) were developed based on questions related to each element of the CATWOE for their system and subsystems. One possible RD for the total system of one company was:

A system owned by the Board of Directors and administered by a Managing Director with the purpose of converting supermarket requirements into the decisions and actions necessary to consistently deliver fresh produce at the desired times, quantities, and qualities and at margins lower than the costs for supermarkets to internalize the operations given that margins are limited and it is difficult to source fresh produce.

Over 15 Conceptual Models (CM) at various levels of detail were developed from the RDs. The total system CM had activities for administer, support sales, trade, ensure quality, and handle product. The CM for the *trade* subsystem included sub-subsystems for the activities: buy, sell, and manage. One possible RD for the *trade* subsystem was:

A system owned and operated by the trade managers for the purpose of establishing and maintaining long-term relationships with senders and supermarkets while constrained by the need to make a profit.

SSM provided a framework for learning about the operations of importing companies through the development of root definitions and conceptual models. The ability to combine the input from several managers into a CM of the total system was a notable contribution of SSM. Working as a team was critical to successfully documenting our learning about importing companies. One company used results from the SSM study to analyze the subsystems that would need to be added or changed and decided the timing was not favorable for implementing ISO 9000 Quality Management Programs.

Kuwait

In 1992 and 1993 multiple consulting assignments that included use of SSM were completed for the Kuwait Institute for Scientific Research (KISR). One project was to develop a proposal for the Food Technology Group at KISR to obtain internal funding to conduct research for reducing postharvest losses (Cassens et al., 1993). SSM was used to gain an understanding of the businesses, institutions, and agencies involved with food production, processing, distribution, and consumption in Kuwait. Then root definitions and conceptual models were developed for tomato production and marketing to demonstrate applications of SSM as described in the project proposal.

The second project for KISR (Prussia, 1993) was part of a larger project to develop a strategic plan for food security in Kuwait. A comprehensive conceptual model was developed for visualizing the interactions of the businesses and agencies involved with the production, importation, and distribution of food along with its consumption. The results helped to clarify the role of government agencies in gaining and providing knowledge to businesses.

One specific result of the second project was the concept of locating/building receiving facilities at the two main border crossings with Saudi Arabia. Trucks with food and other products from shipping countries could be unloaded and the cargo reloaded onto Kuwaiti vehicles. At the time of the assignment all trucks had to stop along the roadway at the border for several separate inspections (military, immigration, customs, and quality standards). In addition, security considerations required all trucks arriving from other countries to wait at the border for a daily armed convoy to warehouses in Kuwait City.

Facilities at the border would help improve the quality of fresh produce by reducing the number of times the doors would need to be opened on refrigerated cargo. Cross-docking at refrigerated docks would maintain the cool chain. Sample boxes removed for quality evaluations would also be more representative of the shipment because they could be selected at random from all parts of the load as the cargo was transferred.

Mexico

A collaboration agreement between the University of Georgia and the University of Veracruz in Xalapa, Mexico, included a workshop in Xalapa for training on SSM (Prussia et al., 1995). Participants gained an operating knowledge of SSM in less than two days. Four teams were assembled to represent mango producing and harvesting, packing, marketing, and retailing. A short lecture was given to all participants on the next stage of SSM before the four teams developed material for that stage. Each team presented results for their part of the chain before the lecture on the next stage.

A second exercise was to develop possible organizational structures for a new extension program for fruit and vegetable production and marketing in the state of Veracruz. The group was rearranged into two teams using SSM. One approached the project from the perspective of growers and packers and the other team represented brokers and retailers. The results from the first team was a traditional university-based extension organization while the second team decided that the extension system should be owned by the distribution end of the food chain.

Brazil

A visiting scientist from Brazil studied SSM at the University of Georgia for the purpose of learning ways to improve fresh fruit and vegetable businesses in Brazil. A root definition was developed for exporting associations that included the following CATWOE:

Customer—agribusiness managers

Actors—members of the exporting association

Transformation—dynamically improve concepts for efficient postharvest handling of horticultural products and prepare dynamic procedures for each business

World view—a system can be developed for enhancing Marketing Quality Life (MQL) for target markets

Owner—executives of the exporting association

Environment—reluctance to new management systems. Managers may not want to be told how to improve their business, although it is necessary to do so for reaching the desired goal(s) of target markets.

When a conceptual model was developed for a Market Coordination System, it was realized that an activity for *learning* was a subsystem that would have been the most likely to overlook when designing an organizational structure. A subsequent consulting assignment in Brazil was completed for the Table Grape Exporting Board (Prussia, 1995). A result was the development of a prototype quality management program for packinghouses that was patterned after ISO 9000 guidelines.

REFERENCES

Campbell, D. T. 1985. A Systems Analysis of Postharvest Damages to Fresh Fruits and Vegetables. M.S. Thesis, Univ. of Georgia, Athens.

Cassens, R. G., Prussia, S. E., and Wilson, G. 1993. Reduction in postharvest and post mortem losses/spoilage in perishable foods in the state of Kuwait. Consultant's report for the Kuwait Institute for Scientific Research. Project proposal, 123 pp.

Checkland, P. 1993. *Systems Thinking, Systems Practice,* John Wiley & Sons, Chichester, England.

Checkland, P. and Scholes, J. 1990. *Soft Systems Methodology in Action,* John Wiley & Sons, Chichester, England.

Kantor, L. S., Lipton, K., Manchester, A., and Oliveira, V. 1997. Estimating and addressing America's food losses. *Food Review,* Jan-April, 2–12.

Okezie, B. O. 1998. World food security: the role of postharvest technology. *FoodTechnology,* 52 (1) 64–69.

Prussia, S. E. 1993. Postharvest handling needs in Kuwait. Consultant's report to the Kuwait Institute for Scientific Research, Strategic Plan for Task 2.A of Kuwait's agricultural planning document AG-67. 68 pp.

Prussia, S. E. 1995. Dynamic technological management for enhancing the postharvest quality of tropical horticultural products: table grape systems. Consultant's final report to EMBRAPACPATSA, Petrolina-PE, Brazil. 15 pp.

Prussia, S. E., Beristain, C. I., Rohs, F. R., and Cortes, J. 1995. Postharvest handling: Implementing a systems approach in Mexico, in *Proceedings of the International Conference on Harvest and Postharvest Technologies for Fresh Fruits and Vegetables,* Guanajuato, Mexico, 20–24 February. pp. 106–112.

Prussia, S. E. and Hubbert, C. R. 1992. Soft system methodologies for analyzing international produce marketing systems. *Acta Horticulturae,* 297: 649–654

Prussia, S. E. and Shewfelt, R. L. 1993. A systems approach to postharvest handling, in *Postharvest Handling: A Systems Approach,* R. L. Shewfelt and S. E. Prussia, eds., Chapter 3, Academic Press, Orlando, FL.

Shewfelt, R. L. and Prussia, S. E. (ed.) 1993. *Postharvest Handling: A Systems Approach,* Academic Press, Inc., San Diego, CA.

Watada, W. A. 1987. Vitamins, Ch. 22, in *Postharvest Physiology of Vegetables,* J. Weichmann, ed., Marcel Dekker, Inc., New York.

Wilson, B. 1990. *Systems: Concepts, Methodologies, and Applications,* John Wiley & Sons, Chichester, England.

Wilson, K. and Morren, G. E., Jr. (Ed.) 1990. *Systems Approaches for Improvement in Agriculture and Resource Management,* Macmillan Publishing Co., New York.

CHAPTER 15

Integrating Problem-Oriented Research

ROLF KUCHENBUCH
BERNHARD BRÜCKNER
JÖRG RÜHLMANN
BARBARA RÖGER

INTRODUCTION

IN this chapter we will focus on research that aims at problem solving in the area of vegetable and fruit quality. This may include basic research, applied research, and development. But all levels of research aim at solving existing problems. Three questions have to be answered: what are the expectations (problems), whose expectations (problems) are considered relevant and how can research help to meet the expectations. (e.g., solving the problems)? Since we want to explain and illustrate our vision we use the approach of the Institute of Vegetable and Ornamental Crops as an example.

WHAT ARE THE PROBLEMS?

The targets of the agricultural and horticultural sector are not static, but have been modified in the past (Carlsson, 1995). Involvement of society followed different objectives as ensuring supply, increasing productivity, lowering cost of food or shifting labor force into most productive sectors of society.

Changes in the last two decades often originated from outside the agricultural sector. The orientation included sustainable agriculture, reduced pesticide use and rural development. General debates on values in society had an influence on the political system, which stimulated change into a politically desired direction. In some countries farmer organizations adopted new points of view, formerly not associated with knowledge of agricultural production. In some cases this also had marketing reasons.

Horticultural production in Germany today, and to some degree all over the world, is characterized by a target system that may be described by the interaction of the components yield (economy), environment, and quality (Figure 15.1). The problems arising from these interactions are as follows.

High *yield* per se is no longer the main aim of horticultural production, even though it is the basis for a sound economy of the enterprises considered. The German market is characterized by imports that fill the gaps of supply from domestic production, only slightly increasing consumption, keeping stagnant or even decreasing retail prices. In a situation like this product attributes other than the price give relative advantages for one product over the other.

Integration
Target System

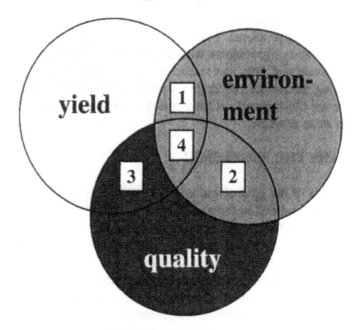

FIGURE 15.1 Target system.

Increasing concern is focused on the *environmental* impact of horticultural production that is characterized by a high input of fertilizers, pesticides, and in some cases heavy machinery. It is obvious and has been demonstrated excessively that consistently high and economical yields are very difficult to achieve if the input of these factors is reduced. Hence, the overlap between yield and environment, area "1" in Figure 15.1, is to be considered as a problem area that must be covered by research.

Including *quality* into the target system complicates matters even more. This is due to the fact that all three targets are interdependent when one considers the production and handling procedures that are currently used to achieve these goals. Factor input to achieve high internal and external quality of the product, area "2" in Figure 15.1, is a key area of future research, whereas we feel that that optimization of yield and quality without consideration or even at the expense of the environment, area "3" in Figure 15.1, will lose importance.

The challenge for the future research is area "4" in Figure 15.1: integrating the production goals yield, environment, and quality. In this field we have the least information at the moment, and research is complex, timely and expensive. This fact calls for new approaches.

One example of how the competing goals within the horticultural production can be approached comprises the supply of mineral and organic fertilizer to improve yield and quality but also to protect natural resources and avoid losses to the environment. The application of nitrogen using organic and mineral fertilizers in agriculture is necessary to balance the amounts removed by harvested produce and waste. The uptake itself can be controlled within a wide range by the amount of fertilizers applied.

Six long-term experiments, lasting 12–28 years and performed on sandy and loamy soils were analyzed for the balance of supplied, used and lost quantities of nitrogen fertilizers (Rühlmann and Geyer, 1999). The supplied nitrogen clearly increased the amount of nitrogen taken up by the canopy (Figure 15.2). The effectiveness of mineral and organic sources of nitrogen to increase the nitrogen uptake by the plants was different for the mineral and organic fertilizers; e.g., 55% of the applied mineral nitrogen and only 15% of the applied organic nitrogen was used by the plant. The remaining part was either used to build up organic nitrogen fractions in the soil or was lost either to the atmosphere or to the groundwater. Strategies that avoided an average annual nitrogen supply above 120 kg ha^{-1} led to average annual losses of less than 10 kg ha^{-1}.

N-uptake [kg/ha]

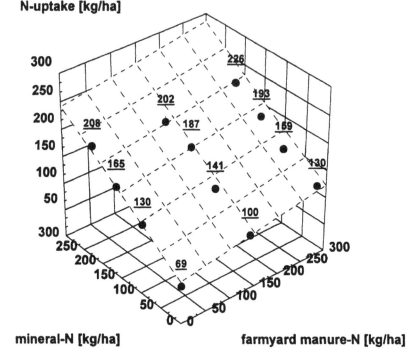

mineral-N [kg/ha]   **farmyard manure-N [kg/ha]**

FIGURE 15.2 Effect of nitrogen supply from mineral fertilizer and farmyard manure on the nitrogen uptake of vegetable crops. Institute for Vegetable and Ornamental Crops, Grossbeeren, Germany, means 1989–1997.

Strategies that obeyed this threshold value on a long-term scale can contribute to avoid nitrogen losses of agro-ecosystems.

To evaluate these fertilizer strategies, their effects on total yield and produce quality have to be taken into account. Therefore a long-term box plot experiment was carried out by the Grossbeeren Institute. During the 24-year period of the experiment three levels of mineral nitrogen fertilizer were combined with different organic fertilizers that were applied annually to each of the plots.

In 1998 carrots ('Nanthya F1') were harvested from those plots in July. They received 0 kg N/ha, 40 kg N/ha from organic (farmyard manure), 120 kg N/ha from mineral or 160 kg N/ha from a combination of organic and mineral fertilizers. The increase of N-fertilization led to a significant linear increase of the yield, and some increase in the enzymatically measured sugar content was recorded (Table 15.1). The percentage of juice extractable by a household press did not change significantly.

Table 15.1. *The Results of Different Nitrogen Fertilization on the Estimated N-uptake, the Recorded Yield and Sugar Contents of Carrot (var. 'Nanthya F1').*

N-Input	Estimated Uptake (kg/ha)	N-Yield (t/ha)	Sucrose (mg g^{-1})	Total Sugar (mg g^{-1})	Juice (mg g^{-1})
0	78	20.4 a	23.2 a	50.6 a	45.7 a
40 kg/ha organic	99	30.2 ab	26.5 ab	52.1 a	46.9 a
120 kg/ha mineral	144	35.2 ab	28.2 b	51.4 a	49.4 a
160 kg/ha organic and mineral	165	42.4 b	28.7 b	53.9 a	49.6 a

From a consumer-oriented standpoint the measurement of few physical attributes may be not sufficient to estimate the acceptance of the product. Therefore sensory acceptance tests with 100 selected, but untrained housewives were conducted. The raw carrots were presented in randomized order and monadically. The consumers marked their degree of acceptance on an unstructured line scale. The liking of the "first impression" was influenced by and correlated with the liking of the shape and color, whereas the "overall liking" reflected, in addition, the liking of mouthfeel, taste and flavor properties.

A significant increase of the consumer acceptance values occurred when nitrogen fertilizer was supplied (Figure 15.3). Differences between the three levels of fertilization were not significant for "overall liking," but for "first impression." The maximum value of "overall liking" was reached at the 40 kg N/ha level, the maximum value for the liking of the "first impression" was reached at the 120 kg N/ha level, significantly different from the other levels. From this consumer standpoint the optimal internal and external quality could be reached within the 40 and the 120 kg N/ha fertilization. On a long-term scale this would not conflict

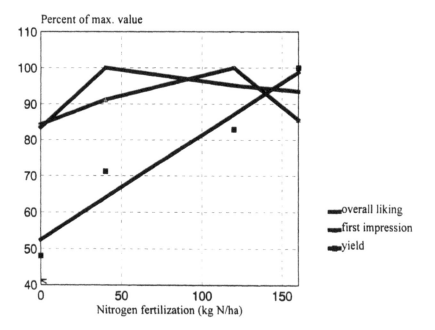

FIGURE 15.3 Relative yield of carrots ('Nanthya F1') and consumer acceptance values of "first impression" and "overall liking" as dependent on nitrogen fertilization.

with the avoidance of nitrogen losses to the environment. The yield may, however, be reduced, in our example by 20%.

Even if the economic, quality and ecological goals of horticultural production do not remain conflict-free, the interdisciplinary research approach as shown in the above example offers the basis for balancing seemingly contradicting values.

WHOSE PROBLEMS HAVE TO BE SOLVED?

Integrating the production goals yield, environment, and quality is of interest for several interest groups who may be considered supporters or stakeholders of research in the field of vegetables and fruit. These are depicted in Figure 15.4 and connected by lines, indicating that they are not acting independently. For each problem, one set of interconnecting

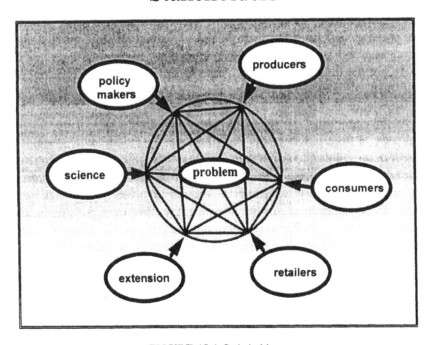

FIGURE 15.4 Stakeholders.

lines will be valid. However, these stakeholders look at the target system very differently, e.g., they estimate problems differently.

One very simple, but unsolved set of problems associated with the production of spinach may illustrate this situation. Spinach is fast growing, has a high demand for nitrogen and is harvested in full growth.

Viewpoint of the *consumer* interested in environmental questions and *policymakers*: Fertilization practices imply a high input of nitrogen fertilizer and as a consequence high nitrate content of spinach leaves and more nitrogen to be left in the soil after harvest than acceptable under environmental considerations. Both may be considered undesirable.

Viewpoint of the *producer*: Two measures can be taken: reduction of fertilizer input and/or planting of cash crops after harvest. Reduced fertilization reduces yield appreciably and planting of a cash crop may increase production cost.

Viewpoint of the *consumer*: May be mainly interested in high-quality spinach without being too conscious when it comes to fertilizer application rates, pesticide use, and so on. Intrinsic factors of quality are the main focus.

Research that is oriented toward optimization of a production system (like the example of spinach) must in our opinion show the impacts for the different groups of stakeholders. This is possible by looking at both the problem and its solution from the viewpoint of different groups. Doing that will increase the understanding that "one cannot have it all." Modeling the interdependencies mathematically may serve as a tool to investigate alternative production strategies or products and find out gaps of knowledge that have to be filled.

At the same time, we face a situation where reduction of the financial support from governments has raised the discussion of "who should pay" for agricultural research and extension and of the possibility to find market solutions. These considerations will include the wish to control the activities by those stakeholders who are involved in financial support. A consequence might be an orientation toward a punctual optimization of actual problems through regional solutions while neglecting the integrative potential of a basic and problem-oriented research.

HOW CAN THE PROBLEMS BE SOLVED?

It is common sense that different problems need different approaches for their solution. But it is evident from the above arguments that not only must the target system and the viewpoints of the stakeholders be integrated as much as possible, but it is also inevitable that scientific disciplines be integrated to solve the problems. This, in our opinion, is a prerequisite for problem-oriented and useful research in a complex environment.

Traditionally agricultural research developed by adopting the methodology of the basic natural sciences such as chemistry, physics, botany and others. With expanding knowledge and increasing differentiation, the "agricultural disciplines" began to develop more or less independently of each other (Krug, 1982). The specialized disciplines were very successful in utilizing the progress originating from basic science. This knowledge was on a low abstraction level, allowing for effective analysis of cause/effect relationship. On the other hand risks of reductionism are introduced when crop growth control, including the postharvest interface, rely only on a theoretical combination of even well-understood processes.

These drawbacks have been recognized and have led to efforts to regard the system as a whole and integrate all factors involved until harvest on a high level of abstraction. The crop was considered as a black box, reactions (growth, yield) were modeled as a function of the inputs. To make the resulting crop growth models robust to changing and unexpected circumstances, knowledge of the mathematical structure of the crop behavior was successively included.

The situation becomes additionally complicated, when the investigation of influences before harvest combined with those during and after harvest is extended to the point of consumption. This approach opens the view of the whole system including operations from breeding to consumer preference. The inevitable consequence of this more system-oriented approach is the integration of research disciplines, which in itself could favor organizational optimizations of the research units.

Figure 15.5 depicts this approach using the structure of the Institute for Vegetable and Ornamental Crops as an example. The Institute itself has competence in the fields of plant nutrition, plant health, quality, modeling/transfer of knowledge, and plant propagation. Since problems do not define themselves according to scientific disciplines, and mostly are not exclusively important for one stakeholder, the approach to problem solving should be interdisciplinary and/or transdisciplinary. Interdisciplinarity in that respect means that highly specialized, well-trained scientists gather around a problem and try to solve it by practicing horizontal integration of knowledge.

The traditional flow of knowledge from basic research through ap-

Integration
Disciplinary Competence

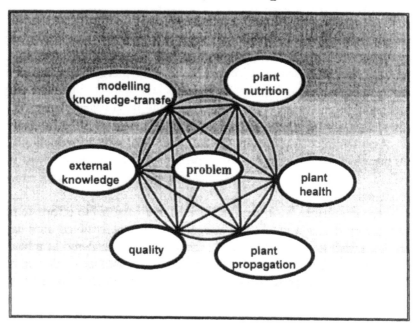

FIGURE 15.5 Disciplinary competence.

plied research, development, extension to practical growers was structured into almost independent steps, separated by time and distance. Transdisciplinarity implies that vertical integration of the research process takes place, e.g., from basic research to development of, e.g., cultural practices at the same time to make rapid use of the research results possible.

Because the problems are complex and no one research institute covers all the fields necessary for problem solving, cooperations and strategic alliances via the formation of networks to make use of external knowledge (Figure 15.5) are the way to successfully solve problems.

REFERENCES

Carlsson, M. 1995. On Agricultural Knowledge Systems. Some Reflections Based on Country Notes and Case Studies. *First Joint Conference of Directors and*

Representatives of Agricultural Research, Agricultural Advisory Services and Higher Education in Agriculture, AGR/REE(95)4. OECD Headquarters, Paris.

Krug, H. 1982. Methodology trends in horticultural research. *Proceedings of the 21ˢᵗ International Horticultural Congress*, Hamburg, pp. 1218–1223.

Rühlmann, J. and Geyer, B. 1999. N-Bilanzen in Langzeitversuchen mit Gemüse. 5. Wissenschaftstagung zum Ökologischen Landbau "Vom Rand zur Mitte," 23.–25.02.1999, Berlin, Posterpräsentation.

CHAPTER 16

A More Integrated View

ROBERT L. SHEWFELT
LEOPOLD M. M. TIJSKENS

INTRODUCTION

THIS book is a product of an international conference held in Potsdam to look for greater integration of fruit and vegetable research from crop production to postharvest physiology to quality evaluation to economics. The general conclusions of the conference were that there are many ways to view fresh fruits and vegetables, that research efforts could be enhanced through greater integration of approaches, but that there are many barriers to greater integration. One of the primary barriers is that of terminology, particularly with respect to quality and acceptability. The problem appears not to be a lack of common terms, but rather that the same terms are used from many perspectives but with important differences in meaning. This chapter seeks to bring together different perspectives on these topics to highlight areas of common ground, describe points of contention and propose a series of guidelines that would facilitate greater integration between studies and approaches.

QUALITY

Many definitions of quality appear in this book as shown in Table 16.1. The standard definitions of fitness for use (Juran, 1974, cited by Tijskens, 1999) or purpose (Simmonds, 1979, cited by Pecher and von Oppen, 1999) and conformance to requirements (Crosby, 1979, cited by

Table 16.1. Definitions of Quality Used by Chapter Authors in This Book.

Chapter	Definition	Source
1	fitness for use	Juran, 1974
	meeting the expectations of the consumer	Simmonds, 1979
3	fitness for purpose	Simmonds, 1979
8	composite of those characteristics that differentiate individual units of a product and have significance in determining the degree of acceptability of that unit by the buyer	Kramer and Twigg, 1970
11	exceeding the expectations of the customer	Oliver, 1997
	products and services that satisfy or exceed customers' requirements	Surak and McAnelly, 1992
12	composite attribute of products that have economic or aesthetic value to the user	Hill, 1996
13	continuous improvement	Deming, 1986
	conformance to requirements	Crosby, 1979
	[a] product that is most economical, most useful and always satisfactory to the consumer	Ishikawa, 1985

Viaene et al., 1999) narrow consideration of quality to the absence of defects and do not permit consideration of degree of excellence. These definitions are not considered sufficient for fresh fruits and vegetables, which deteriorate during storage. In addition, flavor quality is not merely an absence of off-flavors but the presence of differing levels of aromatic components (related to flavor excellence). The importance of the consumer is incorporated into other definitions that relate to their expectations (Surak and McAnelly, 1992, cited by Bech, 1999; Jongen, 1999; Mazur, 1987, cited by Bech, 1999; Kramer and Twigg, 1970, cited by Shewfelt, 1999). Economics provides the basis for other definitions (Hill, 1996, cited by Florkowski, 1999; Ishikawa et al., 1985, cited by Viaene et al., 1999). Finally, quality is also described as continuous improvement (Deming, 1986, cited by Viaene et al., 1999).

Adoption of a single definition for quality would reduce the difficulty in comparing studies from different laboratories and would foster greater collaboration across disciplines. Any definition must be broad enough to encompass a wide range of perspectives but narrow enough to be meaningful. Such a definition of quality must be grounded in product characteristics that are readily measurable and can be translated into terms of consumer acceptability and economics. For practical applicability within the framework of available knowledge and skills, the focus should be on obtaining readily measurable characteristics that are meaningful across all links in the postharvest chain.

RELATED TERMS

To understand quality other terms must be considered. Definitions used in other chapters in the book are shown in Table 16.2. Quality characteristics are "measurable technical attributes" (Hauser and Clausing, 1988, cited by Bech, 1999). Quality can be either intrinsic—associated with the product or extrinsic—associated with perception of the product within the handling system (Jongen, 1999; Shewfelt, 1999; Tijskens, 1999). Intrinsic quality can be influenced by maturity of the item at harvest (Nilsson, 1999). Customers are any users in the whole chain from end user to seed developer (Bech, 1999).

Two terms, "acceptability" and "acceptance," relate to the extrinsic factors of quality. Even though both terms describe similar concepts, there are subtle differences between them (Tijskens, 1999). Acceptability implies that some criterion is applied to differentiate those products that can be sold from those that cannot, while acceptance does not imply

Table 16.2. *Definitions of Terms Related to Quality Used by Chapter Authors in This Book.*

Term	Definition	Chapter(s)/ Source
Acceptability	implies that some criterion is applied to differentiate those products that can be sold from those that cannot	7
	the level of continued purchase or consumption by a specified population	8/Land, 1988
Acceptance	does not imply any aspect involving a range in quality in the general state of a product	7
	actual utilization (purchase, eating) may be measured by preference or liking for a specific item	
	"an experience or feature of an experience, characterized by a positive (approach in a pleasant) attitude"	10/Amerine et al., 1965, p. 549
Intrinsic quality	associated with the product	1, 7 and 8
Extrinsic quality	associated with perception of the product within the handling system	1, 7 and 8
Keeping quality	time the product's quality remains acceptable during storage and transport	7
Quality characteristics	measurable technical attributes	11/Hauser and Clausing, 1988
Shelf life	the time period a product can be expected to maintain a predetermined level of quality under specified storage conditions	8

any aspect involving a range in quality in the general state of a product (Tijskens, 1999). Thus, acceptance describes the state of a product, while acceptability describes the interaction of that product with the consumer and existing conditions. Other definitions of acceptability include "the level of continued purchase or consumption by a specified population" (Land, 1988, p. 476, cited by Shewfelt, 1999) and "actual utilization (purchase, eating) may be measured by preference or liking for a specific item" or "an experience or feature of an experience, characterized by a positive (approach in a pleasant) attitude" (Amerine et al., 1965, p. 543, cited by Brückner and Auerswald, 1999).

APPROACHES

Likewise numerous approaches to the study of quality or application of basic principles of quality have been described, as listed in Table 16.3. Specific models that have been advanced range from the hard, clearly defined outputs of Quality Function Deployment (Bech, 1999) to the soft, more flexible outputs of Soft Systems Management (Prussia, 1999). These models incorporate rather generalized concepts like Total Quality Management and Efficient Consumer Response to very specific applications such as Integrated Quality Management (Viaene et al., 1999). Product distribution is emphasized in Supply Chain Management (Jongen, 1999; Viaene et al., 1999), while consumer response is critical in Quality Enhancement (Shewfelt, 1999) and the Total Food Quality Model (Bech, 1999). Quality has been viewed from the standpoint of the product with respect to gene-technological modification (Wehling, 1999), maturity (Nilsson, 1999) and grades and standards (Florkowski, 1999). It has been viewed from the standpoint of the consumer with respect to consumer demand (Pecher and von Oppen, 1999), food choice (Adani and MacFie, 1999) and quality limit, which integrates the concepts of quality and acceptability (Tijskens, 1999). Quality has also been viewed from the standpoint of market valuation (Florkowski, 1999). Measurement of quality has been related to the objective/subjective framework of a product orientation (Pecher and von Oppen, 1999; Bech, 1999) and to the purchase/consumption of a consumer orientation (Shewfelt, 1999; Brückner and Auerswald, 1999; Florkowski, 1999). When expectations do not match reality, stakeholders experience problems, and it is these problems that can serve as a point of integration for research (Kuchenbuch et al., 1999).

Approach	Description	Chapter(s)
Consumer demand	consumer choices seek to maximize utility under constraints of a limited budget	3
Efficient consumer response	combines both supply-side and demand-side elements of Supply Chain Management	13
Food choice	determinants that affect consumer attitudes that influence purchase and consumption decisions	2
Gene-technological modification	modify quality characteristics by manipulation of genetic material	2
Grades and standards	governmental classification of quality that can serve as a basis for pricing	12
Integrated quality management	integration of individual Total Quality Management concepts across the supply chain	13
Market valuation	the market establishes the value of a particular item	12
Maturity	stage of development of an item as affected by physiological and environmental and commercial factors at harvest and during distribution	6
Objective/subjective	measurable technical attributes/consumer sensory perception	3 & 11
Purchase vs. consumption quality	differing criteria for selection and eating	8, 10, & 12
Problem-oriented research	focusing research on areas where reality does not meet expectations	15
Quality enhancement	expresses consumer acceptability as a function of critical quality attributes	8
Quality function deployment	translation of consumer wants and needs into quality characteristics	11
Quality limit	criterion a consumer applies to quality relative to acceptability	7
Soft systems methodology	systems thinking necessary to define existing situations, identify changes needed and take actions for improvement	14
Supply chain management	focuses on linkages in a chain from primary producer to end consumer	1 & 13
Total food quality model	overall satisfaction/dissatisfaction as a function of meeting or failing to meet expectations	11
Total Quality Management	designed to continually improve quality of products and services	13

COMMON GROUND

A view of quality has emerged from the pages of this book. Quality is not a simple concept. Rather, it can be viewed from many perspectives including the product, the consumer and the market. Classification of quality characteristics in fruits and vegetables requires an understanding of both the visible (external, purchase) and hidden (internal, consumption) aspects as well as those associated with the item (intrinsic, sensory) and other factors (extrinsic, image). The importance of specific characteristics changes as the fruit or vegetable travels in the distribution chain and the perspective of a specific customer. The perishable nature of fresh fruits and vegetables ensures that quality is not constant but changes with time, thus introducing the topic of shelf life or keeping quality. Quantitative measures for quality characteristics are required for studies designed to improve the quality of fresh fruits and vegetables. Economic aspects could represent a unifying factor as an item proceeds from harvest to consumption as long as value can be separated from marketing and merchandising considerations.

DIVERGENCE

Many areas are described in this book regarding quality on which there is no consensus. This situation can be viewed in two ways: as an impediment to further progress or an opportunity for coalescing different perspectives. Several definitions of quality were advanced (Table 16.1), but no single definition emerged as superior. The focal point of emphasis ranged from the product to the consumer to the market, with associated terminology related to the focus adopted by chapter authors (Table 16.2). Economic and physiological aspects were either ignored or were considered to be the most important factors in determining quality. Likewise, the importance of shelf life tended to be either overlooked or strongly emphasized. Few quantitative measures of quality or related terms were provided with no consistency between offerings.

A primary reason for this divergence appears to be a confusion between the pure intrinsic quality of a product and its practical acceptability to the consumer or user. Intrinsic quality is completely defined by those properties in the product that have a bearing on the quality characteristics. Acceptability is the result of an evaluation by the user of those quality characteristics in view of an intention of use in a framework that encompasses sociological, psychological, and economic considerations (Steenkamp, 1989; Sloof et al., 1996).

RECOMMENDATIONS

While integration of fruit and vegetable studies requires a more consistent set of terms, too rigid standardization could stifle creativity and innovation. We make the following recommendations to achieve both consistency and flexibility:

1. Consider viewing quality characteristics may be viewed from the three perspectives: intrinsic product properties (e.g., amount of pectin), measurable quality attributes (e.g., firmness or texture), and evaluation by the user (e.g., preference for firm apples).
2. Focus on "measurable technical attributes" of the product when describing quality characteristics.
3. Define quality as the composite of product characteristics that impart value to the buyer or consumer (merges the definition of Hill, 1996, with that of Kramer and Twigg, 1970).
4. Define keeping quality as "the time a product's quality remains acceptable during storage and transport" under dynamic changing conditions (Tijskens, 1999).
5. Define shelf life as the time period that quality remains acceptable to the consumer within retail distribution and home storage under normal storage conditions (merges definitions of Tijskens, 1999, and Shewfelt, 1985).
6. Define acceptability as willingness to purchase or consume a product within a given target population (merges the definitions of Land, 1988, with Brückner and Auerswald, 1999). It could be expressed as a percentage distribution of that population (Shewfelt et al., 1997).
7. Define quality limit as the level of a quality characteristic or product property required to maintain acceptability (adopted from Tijskens, 1999).
8. Avoid in scientific studies the use of general words like "quality" without adequate definition. Precise use of quality characteristics, product properties and consumer acceptability is preferred.
9. Use the term "meeting expectations" (Adani and MacFie, 1999) to describe the reaction of consumers after product evaluation as it relates to acceptability before evaluation.

These recommendations provide terms that encompass product, consumer and market orientations to quality. They confine the measurement of quality to the characteristics of the fruit or vegetable. These characteristics must be identified and can be reported individually or as a composite. "Acceptability" would be the term reserved for understanding consumer response. Quality limit links quality and acceptability. Economic factors are implicit in all definitions. Collection of economic data in quality improvement studies, while preferable, is not required. The concept of keeping quality provides a means of incorporating physiological changes and storage stability into these studies. All terms described are readily quantifiable.

REFERENCES

Adani, Z. and MacFie, H. J. H. 1999. Consumer preference, in *Fruit and Vegetable Quality: An Integrated Approach,* R. L. Shewfelt and B. Brückner, eds., Technomic Publishing Co., Inc., Lancaster, PA.

Amerine, M. A., Pangborn, R. B. and Roessler, E. B. 1965. Physical and chemical tests related to sensory properties of foods, in *Principles of Sensory Evaluation of Food,* M. L. Anson, C. O. Chichseter, E. M. Mrak, and G. F. Stewart, eds., Academic Press, New York.

Bech, A. C. 1999. House of quality—and integrated view of fruit and vegetable quality, in *Fruit and Vegetable Quality: An Integrated Approach,* R. L. Shewfelt and B. Brückner, eds., Technomic Publishing Co., Inc., Lancaster, PA.

Brückner, B. and Auerswald, B. 1999. Instumental data—consumer acceptance, in *Fruit and Vegetable Quality: An Integrated Approach,* R. L. Shewfelt and B. Brückner, eds., Technomic Publishing Co., Inc., Lancaster, PA.

Crosby, P. B. 1979. *Quality Is Free: The Art of Making Quality Certain,* McGraw-Hill, Inc. New York.

Deming, W. E. 1986. *Out of the Crisis,* MIT Center for Advanced Engineering Study, Cambridge MA.

Florkowski, W. J. 1999. Economics of quality, in *Fruit and Vegetable Quality: An Integrated Approach,* R. L. Shewfelt and B. Brückner, eds., Technomic Publishing Co., Inc., Lancaster, PA.

Hauser, J. R. and Clausing, D. 1988. The house of quality. *Harvard Business Review,* 66(2): 66–73.

Hill, L. D. 1996. Quality in agricultural products, in *Quality of U.S. Agricultural Products,* L. D. Hill, eds., Task Force Report No. 126, pp. 13–18. Council for Agricultural Science and Technology.

Ishikawa, K. 1985. *What Is Total Quality Control? The Japanese Way,* Prentice Hall, Englewood Cliffs, NJ.

Jongen, W. M. F. 1999. Food supply chains: from productivity towards quality, in *Fruit and Vegetable Quality: An Integrated Approach*, R. L. Shewfelt and B. Brückner, eds., Technomic Publishing Co., Inc., Lancaster, PA.

Juran, J. M. 1974. Basic concepts, in *Quality Control Handbook*, 3rd Edition, J. M. Juran, F. M. Gryna, and R. S. Bingham, eds., McGraw-Hill, New York.

Kramer, A. and Twigg, B. A. 1970. *Quality Control for the Food Industry*, 3rd edition. Vol. 1, AVI/ Chapman and Hall. London.

Kuchenbuch, R., Brückner, B., Rühlmann, J. and Röger, B. 1999. Integrating problem-oriented research, in *Fruit and Vegetable Quality: An Integrated Approach*, R. L. Shewfelt and B. Brückner, eds., Technomic Publishing Co., Inc., Lancaster, PA.

Land, D. G. 1988. Negative influences on acceptability and their control, in *Food Acceptability*, D. M. H. Thompson, eds., Elsevier, New York, pp. 475–483.

Mazur, G. 1987. *Attractive Quality and Must-Be Quality*. Goal/QPC. Translation of Kano, Seraku, Takashi, Tsuji 1982 from Japanese.

Nilsson, T. 1999. Postharvest handling and storage of vegetables, in *Fruit and Vegetable Quality: An Integrated Approach*, R. L. Shewfelt and B. Brückner, eds., Technomic Publishing Co., Inc., Lancaster, PA.

Oliver, R. L. 1997. *Satisfaction. A Behavioral Perspective on the Consumer*, The McGraw-Hill Companies, Inc., New York.

Pecher, S. and von Oppen, M. 1999. Consumer preferences and breeding goals, in *Fruit and Vegetable Quality: An Integrated Approach*, R. L. Shewfelt and B. Brückner, eds., Technomic Publishing Co., Inc., Lancaster, PA.

Prussia, S. E. 1999. Methods and examples of integration, in *Fruit and Vegetable Quality: An Integrated Approach*, R. L. Shewfelt and B. Brückner, eds., Technomic Publishing Co., Inc., Lancaster, PA.

Shewfelt, R. L. 1985. Postharvest treatment for extending the shelf life of fruits and vegetables. *Food Technol.*, 40(5): 70–72,74,76–78, 80 and 89.

Shewfelt, R. L. 1999. Fruit and vegetable quality, in *Fruit and Vegetable Quality: An Integrated Approach*, R. L. Shewfelt and B. Brückner, eds., Technomic Publishing Co., Inc., Lancaster, PA.

Shewfelt, R. L., Erickson, M. E., Hung, Y-C. and Malundo, T. M. M. 1997. Applying quality concepts in frozen food development. *Food Technol.*, 51(2): 56–59.

Simmonds, N. W. 1979. *Principles of Crop Improvement*, Longman, London.

Sloof, M., Tijskens, L. M. M. and Wilkinson, E. C. 1996. Concepts for modelling the quality of perishable products. *Trends Food Science Tecnol.*, 7: 165–171.

Steenkamp, J. B. E. M. 1989. Product Quality: an Investigation into the Concept and How It Is Perceived by Consumers. Ph.D. thesis, Agricultural University Wageningen, The Netherlands.

Surak, J. G. and McAnelly, J. K. 1992. Educational programs in quality for the food processing industry. *Food Technol.*, 46(6): 88–95.

Tijskens, L. M. M. 1999. Acceptability, in *Fruit and Vegetable Quality: An Integrated Approach*, R. L. Shewfelt and B. Brückner, eds., Technomic Publishing Co., Inc., Lancaster, PA.

Viaene, J., Gellynck, X., and Verbeke, W. 1999. Integrated quality management applied to the processing vegetables industry, in *Fruit and Vegetable Quality: An Integrated Approach*, R. L. Shewfelt and B. Brückner, eds., Technomic Publishing Co., Inc., Lancaster, PA.

Wehling, P. 1999. Quality and breeding—cultivars, genetic engineering, in *Fruit and Vegetable Quality: An Integrated Approach*, R. L. Shewfelt and B. Brückner, eds., Technomic Publishing Co., Inc., Lancaster, PA.

COMMON GROUND

- Traditional disciplinary approaches ignore critical factors in understanding complex interactions during handling and distribution.
- Changes in fruit and vegetable quality must be viewed in the context of handling and distribution.
- Communication between different components of handling and distribution is often inadequate.

DIVERGENCE

- Useful points of integration for research to improve fruit and vegetable quality.
- Theoretical principles and terminology to serve as a focal point of integration.
- The degree of integration needed to provide useful information.

FUTURE DEVELOPMENTS

- A theoretical construct that can serve as a basis of integrated research.
- More integrated studies that focus methodologies from other fields of research on handling of fresh fruit and vegetables.
- A comprehensive integrated model that becomes adopted by commercial firms to improve fresh fruit and vegetable quality.